EAI/Springer Innovations in Communication and Computing

Series editor
Imrich Chlamtac, European Alliance for Innovation, Ghent, Belgium

Editor's Note

The impact of information technologies is creating a new world yet not fully understood. The extent and speed of economic, life style and social changes already perceived in everyday life is hard to estimate without understanding the technological driving forces behind it. This series presents contributed volumes featuring the latest research and development in the various information engineering technologies that play a key role in this process.

The range of topics, focusing primarily on communications and computing engineering include, but are not limited to, wireless networks; mobile communication; design and learning; gaming; interaction; e-health and pervasive healthcare; energy management; smart grids; internet of things; cognitive radio networks; computation; cloud computing; ubiquitous connectivity, and in mode general smart living, smart cities, Internet of Things and more. The series publishes a combination of expanded papers selected from hosted and sponsored European Alliance for Innovation (EAI) conferences that present cutting edge, global research as well as provide new perspectives on traditional related engineering fields. This content, complemented with open calls for contribution of book titles and individual chapters, together maintain Springer's and EAI's high standards of academic excellence. The audience for the books consists of researchers, industry professionals, advanced level students as well as practitioners in related fields of activity include information and communication specialists, security experts, economists, urban planners, doctors, and in general representatives in all those walks of life affected ad contributing to the information revolution.

Indexing: This series is indexed in Scopus and zbMATH.

About EAI

EAI is a grassroots member organization initiated through cooperation between businesses, public, private and government organizations to address the global challenges of Europe's future competitiveness and link the European Research community with its counterparts around the globe. EAI reaches out to hundreds of thousands of individual subscribers on all continents and collaborates with an institutional member base including Fortune 500 companies, government organizations, and educational institutions, provide a free research and innovation platform.

Through its open free membership model EAI promotes a new research and innovation culture based on collaboration, connectivity and recognition of excellence by community.

More information about this series at http://www.springer.com/series/15427

Pandian Vasant • Igor Litvinchev
Jose Antonio Marmolejo-Saucedo
Roman Rodriguez-Aguilar • Felix Martinez-Rios
Editors

Data Analysis and Optimization for Engineering and Computing Problems

Proceedings of the 3rd EAI International Conference on Computer Science and Engineering and Health Services

Springer

EAI
RESEARCH MEETS INNOVATION

Editors
Pandian Vasant
Universiti Teknologi Petronas
Seri Iskandar, Malaysia

Jose Antonio Marmolejo-Saucedo
Universidad Panamericana
Facultad de Ingeniería
Ciudad de México, México

Felix Martinez-Rios
Universidad Panamericana
Facultad de Ingeniería
Ciudad de México, México

Igor Litvinchev
Universidad Autónoma de Nuevo León
San Nicolás de los Garza, Mexico

Roman Rodriguez-Aguilar
Facultad de Ciencias Económicas y
Empresariales
Universidad Panamericana
Ciudad de México, México

ISSN 2522-8595 ISSN 2522-8609 (electronic)
EAI/Springer Innovations in Communication and Computing
ISBN 978-3-030-48148-3 ISBN 978-3-030-48149-0 (eBook)
https://doi.org/10.1007/978-3-030-48149-0

This Springer imprint is published by the registered company Springer Nature Switzerland AG.
The registered company address is: Gewerbestrasse 11, 6330 Cham, Switzerland

Preface

The third edition of the International Conference on Computer Science and Engineering and Health Services (COMPSE) 2019 organized by the Panamerican University took place on November 28 and 29, 2019 in Mexico City. The conference brought together specialists in different fields of engineering and computer science, who contributed their frontier knowledge to the establishment of new challenges to face the future.

The main goal of COMPSE 2019 was to generate pioneer work groups in the development of science in the Latin American region and Asia. This conference offered an opportunity for networking and collaboration among participants. The exchange of ideas as well as peer feedback will benefit future students.

The technical program of COMPSE 2019 consisted of 20 full papers. The conference tracks were: Track 1: Heuristic Design and Optimization; Track 2: Mixed-Integer Programming and Global Optimization; Track 3: Simulation and Distributed Systems; Track 3: Renewable Energies; and Track 4: Information Technologies. Aside from the high-quality technical paper presentations, the technical program also featured three keynote speakers who were Prof. Julio Madera, Prof. Gonzalo Abascal, and Prof. Nathaniel Osgood.

Coordination with the steering chairs, Prof. Imrich Chlamtac, Prof. Pandian Vasant, and Prof. Igor Litvinchev was essential for the success of the conference. We sincerely appreciate their constant support and guidance. It was also a great pleasure to work with such an excellent organizing committee team. In particular, the Technical Program Committee, led by our General Co-Chairs, Prof. Felix Martinez-Rios and Prof. Roman Rodriguez-Aguilar.

We are also grateful to Conference Manager, Karolina Marcinova for her support and all the authors who submitted their papers to the COMPSE 2019. Finally, we sincerely appreciate the support received from the volunteer staff of the Panamerican University who at all times participated in the organization of the event.

Ciudad de México, México
General Chair of COMPSE 2019

Jose Antonio Marmolejo-Saucedo

Conference Organization

Steering Committee	
Imrich Chlamtac	Bruno Kessler Professor, University of Trento, Italy
Pandian Vasant	University Teknologi PETRONAS, Malaysia
Igor Litvinchev	Nuevo Leon State University (UANL), Mexico
Organizing Committee	
General Chair	
Jose Antonio Marmolejo-Saucedo	Universidad Panamericana, Mexico
Felix Martinez-Rios	Universidad Panamericana, Mexico
General Co-Chairs	
Roman Rodriguez-Aguilar	Universidad Panamericana, Mexico
TPC Chair and Co-Chair	
Roman Rodriguez-Aguilar	Universidad Panamericana, Mexico
Workshops Chair	
Javier Aguilar	Universidad Panamericana, Mexico
Publicity & Social Media Chair	
Ernesto Rodríguez	Universidad Panamericana, Mexico
Publications Chair	
Ernesto Rodríguez	Universidad Panamericana, Mexico
Web Chair	
Alfonso Murillo Suárez	Universidad Panamericana, Mexico
Conference Manager	
Karolina Marcinova	EAI
Technical Program Committee	
Jose Antonio Marmolejo-Saucedo	Universidad Panamericana, Mexico
Felix Martinez-Rios	Universidad Panamericana, Mexico
Roman Rodriguez-Aguilar	Universidad Panamericana, Mexico
Ernesto Rodríguez	Universidad Panamericana, Mexico
Alfonso Murillo Suárez	Universidad Panamericana, Mexico
Caridad Gonzalez Sanchez	Universidad Tecnológica de La Habana, Cuba

Diego Fernando Manotas Duque	Universidad del Valle, Colombia
Edith Martínez Delgado	Universidad Tecnológica de La Habana, Cuba
Elsie Noemi Olvera Perez	Universidad de Guadalajara, Mexico
Emilio Jimenez	Universidad de la Rioja, Spain
Giner Alor Hernandez	Instituto Tecnológico de Orizaba, Mexico
Guillermo Mejía Barcena	
Igor Litvinchev	Autonomous University of Nuevo León, Mexico
Ivan Matveev	Russia
Jania Astrid Saucedo Martínez	Autonomous University of Nuevo León, Mexico
Jesus Alfonso Gil Lopez	Universidad Nacional de Educación a Distancia, Spain
Jezreel Mejia Miranda	Centro de Investigación en Matemáticas, Mexico
Jorge Rojas Arce	National Autonomous University of Mexico, Mexico
Leonardo Rivera Cadavid	Universidad del Valle, Colombia
Liliana Avelar Sosa	Universidad Autónoma de Ciudad Juárez, Mexico
Jorge Luis García-Alcaraz	Universidad Autónoma de Ciudad Juárez, Mexico
Julio Blanco Fernandez	Universidad de la Rioja, Spain
Prof. Manuel Velarde	Instituto Tecnologico de Sonora, Mexico
Martin Bogdan	Universität Leipzig, Alemania
Maximiliano Lopez Lopez	National Autonomous University of Mexico, Mexico
Oliverio Cruz-Mejia	Autonomous University of the State of Mexico, Mexico
Pandian Vasant	Universiti Teknologi PETRONAS
Petro Setsyuk	
Raul Figueroa Romero	Universidad Autónoma de México, Mexico
Rigoberto Torres Tovar	
Romeo Sánchez	Autonomous University of Nuevo León, Mexico
Rosario Garza Rios	Universidad Tecnológica de La Habana, Cuba
Socorro Rangel	Universidade Estadual Paulista, Brasil
Tatiana Romanova	
Tomás Eloy Salais-Fierro	Autonomous University of Nuevo León, Mexico
Vladimir Tsurkov	CCAS, Russia
Yolanda Angelica Baez Lopez	Universidad Autónoma de Baja California, Mexico
Zaida Alarcón Bernal	National Autonomous University of Mexico, Mexico
Zoia Duriagina	Ucrania

Contents

Part I Heuristic Design and Optimization

1 A New Heuristic Based on a Parallel Implementation of Firefly Algorithm 3
Alfonso Murillo-Suarez and Felix Martinez-Rios

2 A New Approach of the Rain-Fall Optimization Algorithm Using Parallelization 13
Juan Manuel Guerrero-Valadez and Felix Martinez-Rios

3 An Heuristic Algorithm to Calculate the Minimum Amount of Anesthesia Using MRI 27
Enrique González-Martín, Guillermo Sosa-Gómez, and Omar Rojas

Part II Mixed-Integer Programming and Global Optimization

4 Backbone Distribution Network Design for the Mexican Automotive Industry 41
Brenda Retana-Blanco, Jose Antonio Marmolejo-Saucedo, Roman Rodriguez-Aguilar, and Erika Pedraza-Arroyo

5 A Hybrid Model for Improving the Performance of Basketball Lineups 61
Roman Rodriguez-Aguilar, Rodrigo Infante-Escudero, and Jose Antonio Marmolejo-Saucedo

6 Location of Bases for Prehospital Services 73
Zaida E. Alarcón-Bernal, Ricardo Aceves-García, and Jorge L. Rojas-Arce

7 Balance Layout Problem with the Optimized Distances Between Objects 85
S. Plankovskyy, A. Nikolaev, O. Shypul, I. Litvinchev, A. Pankratov, and T. Romanova

8 **Optimized Packing of Object Clusters with Balancing Conditions** ... 95
T. Romanova, A. Pankratov, I. Litvinchev, and Jose Antonio Marmolejo-Saucedo

Part III Simulation and Distributed Systems

9 **Didactic Tool for Teaching Election Algorithms in Distributed Systems** ... 111
Araceli López-Reyes and Francisco de Asís López Fuentes

10 **Literature Review: Evaluation of the Feasibility of Implementing Industry 4.0 Technologies in the Intralogistic Processes of the Logistics Operators of the Department of the Atlantic, a Look Towards the Continuous Improvement of Organizational Efficiency** ... 125
Jania Astrid Saucedo Martínez and Carlos Regalao Noriega

11 **Secure Key Distribution Prototype Based on Kerberos** ... 143
Abigail Verónica Chantes Barrios and Francisco de Asís López Fuentes

12 **A Decision Making Approach Using Fuzzy Logic and ANFIS: A Retail Study Case** ... 155
Tomas E. Salais-Fierro, Jania Astrid Saucedo Martínez, and Blanca I. Pérez-Pérez

Part IV Renewable Energies

13 **Development of Renewable Energy Resources in Global Energy Interconnection** ... 175
Thu Yein Min, He Haiyang, Michael G. Tyagunov, Tatiana A. Shestopalova, and Aung Ko

Part V Information Technologies

14 **Main Metric Components in the Generation of Mixed Indicators: An Application of SGVD Methodology** ... 195
Roman Rodriguez-Aguilar

15 **SMOTE-Cov: A New Oversampling Method Based on the Covariance Matrix** ... 207
Ireimis Leguen-deVarona, Julio Madera, Yoan Martínez-López, and José Carlos Hernández-Nieto

16 **A Sentiment Analysis Method for Analyzing Users Opinions About Drugs for Chronic Diseases** ... 217
María del Pilar Salas-Zárate, Giner Alor-Hernández, Jorge Luis García-Alcaraz, Luis Omar Colombo-Mendoza, Mario Andrés Paredes-Valverde, and José Luis Sánchez-Cervantes

17 Proofs of the Undecidability of Stegananalysis Techniques 229
Juan Gutierrez-Cardenas

18 Vulnerability of Network Coding Under Pollution Attacks 243
Raúl Antonio Ortega Vallejo and Francisco de Asís López Fuentes

19 An Application of Hadamard Transform to Test Stream Ciphers 255
Guillermo Sosa-Gómez, Omar Rojas, and Octavio Páez-Osuna

20 A Comparison of Speech-to-Speech Neural Network Methodologies for Digit Pronunciation 263
Manuel Quintana and Miguel Bernal

Correction to: Data Analysis and Optimization for Engineering and Computing Problems C1

Index 275

The original version of this book was revised: Contributing Authors "Roman Rodriguez-Aguilar, Brenda Retana-Blanco and Erika Pedraza-Arroyo" affiliations have been updated. The correction to this book is available at https://doi.org/10.1007/978-3-030-48149-0_21

Part I
Heuristic Design and Optimization

Chapter 1
A New Heuristic Based on a Parallel Implementation of Firefly Algorithm

Alfonso Murillo-Suarez and Felix Martinez-Rios

Contents

1.1 Introduction 3
1.2 Parallel Implementation of Firefly Algorithm 5
1.3 Experiments 8
1.4 Conclusions 11
References 12

1.1 Introduction

Nature-Inspired Optimization Algorithms

Continuous optimization problems consist of finding the optimal solution for a function (whether maximum or minimum) in a determined space. Some of these functions have high mathematical complexity and number of variables so that determining the optimum with traditional methods is impossible besides that many computational resources and time would be needed. Even if other methods are used, the complex characteristics of the search space can lead to finding local optima, instead of global [1, 2].

Research about optimization algorithms has increased in recent years due to their need in different fields of science, like engineering, physics, manufacturing processes, economies of scale, allocation and location problems, between others. This has raised the necessity for faster and more accurate algorithms. All this research has led to the development of global optimization algorithms to avoid the problem of getting caught in one of the many local optima a function may have. The design of algorithms that obtain a global optimum is difficult because there is no criterion to determine if a result is local or not [13].

A. Murillo-Suarez (✉) · F. Martinez-Rios
Universidad Panamericana, Facultad de Ingeniería, Ciudad de México, México
e-mail: alfonso.murillosuarez@up.edu.mx; felix.martinez@up.edu.mx

P. Vasant et al. (eds.), *Data Analysis and Optimization for Engineering and Computing Problems*, EAI/Springer Innovations in Communication and Computing,
https://doi.org/10.1007/978-3-030-48149-0_1

Through all the research explained before, some authors have developed algorithms inspired by behaviors found in nature and swarm intelligence. Some examples of this kind of algorithms are the Particle Swarm Optimization (PSO) [4] with lots of variants like the Comprehensive Learning PSO [11] and the Example-Based Learning PSO (ELPSO) [7], Artificial Bee Colony (ABC) [9], Ant Colony Optimization [3], Genetic Algorithms [6], Firefly Algorithm [15], between so much others.

In this chapter, we are focusing on the Firefly Algorithm, which is a barely new algorithm that has proven to obtain good results in a short time. The efficiency of this variant of the Firefly Algorithm allows improving the results of application problems such as the process planning method whose resolution is explained in Sun's paper [14], or the Pressure Vessel and Welded Beam problems, as explained in [10].

Firefly Algorithm

Firefly Algorithm, presented by Dr. Yang [15], is inspired in the mating behavior of fireflies, who are attracted to each other by the intensity of their flashing. The brighter the flashing of a firefly is, the more attractive it is for other fireflies. This light intensity (I) is less perceived as the distance (r) increases because of its absorption in the environment. In the algorithm, this light intensity is formulated according to the position of the firefly in the function's search space.

For this algorithm, three basic rules are proposed:

- The fireflies can be attracted to each other regardless of their sex.
- The attractiveness of a firefly is proportional to their flashing brightness and decreases with the distance between two fireflies.
- The brightness of the firefly is proportional to the value of the objective function in the position on the search space.

Taking these rules into consideration, we can now present the original Firefly Algorithm pseudocode as shown in Algorithm 1 [16].

Algorithm 1 needs some equations to achieve the execution. The first one (Eq. (1.1)) is the position updating equation where β_0 is a constant value that represents the attractiveness when the distance between two fireflies is zero. The term $\beta_0 e^{-\gamma r_{ij}^2}$ determines the loss of attractiveness due to the distance r, which is a Cartesian distance as shown in the second equation (Eq. (1.2)) and the environment absorption (which uses the light absorption coefficient γ).

$$x_i^{t+1} = x_i^t + \beta_0 e^{-\gamma r_{ij}^2} \left(x_j^t - x_i^t \right) + \alpha (random - 0.5) \tag{1.1}$$

Algorithm 1: Firefly Algorithm pseudocode

```
Generate an initial population of fireflies.;
Formulate light intensity (I);
Initialize parameters α, β, γ;
while t < Max_Generations do
    for i = 1 to n do
        for j = 1 to n do
            if I_j > I_i then
                Move firefly i toward firefly j according to Equation 1.1;
            end
        end
        Evaluate new solutions and accept them if better;
    end
    Rank the fireflies and update the best solution found so far;
    Update iteration counter t ← t + 1;
    Reduce α (randomness strength) by a factor;
end
```

For most cases, the value of β_0 is set to 1 and $\alpha \in [0, 1]$.

$$r_{ij} = \sqrt{\sum_{k=1}^{d} \left(x_{i,k} - x_{j,k}\right)^2} \tag{1.2}$$

The alpha parameter varies with each iteration as shown in Eq. (1.3), where α_0 is the initial value of the alpha parameter and θ is the value of the reduction factor in the interval $0 < \theta < 1$.

$$\alpha = \alpha_0 \theta^t \tag{1.3}$$

1.2 Parallel Implementation of Firefly Algorithm

Parallel Algorithms

When talking about a parallel implementation of an algorithm, there are so many advantages. Parallel algorithms were developed as a need to improve the efficiency of processes to obtain results faster. This is possible thanks to the multicore processors available in commercial computers that allow us to run various processes simultaneously [8].

In this chapter, the parallel implementation is performed taking the Firefly Algorithm and executing it in each of the C threads. Each thread starts the execution with independent initial values, which allows the process to have great diversity. In the time in which a simple Firefly Algorithm gives one result, this parallel implementation gives C results, so that we can obtain the best one of these.

Although the idea of a simple parallel implementation gives the opportunity to generate better results, our algorithm presents some techniques to weave the parameters of the threads between them, so that each of them improves with the information of the others after a specific number of iterations, allowing each process to improve itself and, then, improve with the other executions' data.

Later, in section *Restarting techniques*, the proposed methods to interlace the results are explained in detail.

Parallel Firefly Algorithm

For the parallel implementation of Firefly Algorithm, we based in the work published in [12], where a simple procedure is proposed: to run the algorithm in each thread and obtain the better result, but including the restarting methods that we talked about before. The pseudocode for this parallel Firefly Algorithm is shown in Algorithm 2.

Algorithm 2: Modified Firefly Algorithm pseudocode

Initialize C, N, NIT, MAX_NI, k_BEST, S^{best};
for c = *1 to C* **do**
    Initialize $N_c = N, NIT_c = NIT, S_c^{best}$
end
NI = 0;
while $NI \leq MAX\_NI$ **do**
    for *nit = 1 to NIT* **do**
        for c = *1 to C* **do**
            Execute Firefly Algorithm;
            $S_c^{best} \leftarrow$ best value obtained in the execution;
            if S_c^{best} *is better than* S^{best} **then**
                $S^{best} \leftarrow S_c^{best}$
            end
        end
        NI $\leftarrow$ NI + 1;
    end
    $f^k = k\_BEST$;
    foreach c *in* Υ **do**
        Restart the algorithms according to the selected restarting method;
    end
end
Return S^{best}

The variables involved in the execution of Algorithm 2 are described below:

- $C \rightarrow$ number of threads to run the algorithm.
- $N \rightarrow$ number of fireflies.
- $NIT \rightarrow$ maximum number of generations, this value is used to stop the execution of the algorithm for the restarts.
- $MAX\_NI \rightarrow$ the total number of iterations the algorithm will perform.
- $k\_BEST \rightarrow$ number of selected threads to conform the best-threads set.
- $S^{best} \rightarrow$ the best result between all the threads.

In line 3 of the algorithm, the fundamental parameters of each thread are initialized with the same values as the main algorithm, so that they all run with the same basic information.

From line 6 to line 21, the main *while* loop is performed until the maximum number of iterations of the algorithm (MAX_NI) is reached.

In line 7 until the algorithm reaches line 16, a *for* loop is executed to reach the NIT iterations, which is the number of iterations at which the executions in the threads are interrupted to perform the restart.

The *for* loop in line 8 is the one that is parallel performed. Each c of the C threads executes the Firefly Algorithm and registers the best value obtained so far in S_c^{best}, this value is the one that is compared between all the threads to find the best one or the best ones.

Once the executions are interrupted, the k_BEST threads with the best results are sorted from best to worst and registered in f^k, in such a way that we can get the best result or, if needed for the restarting method, the best results.

Restarting Techniques

As it has been mentioned before, the threads interrupt their executions after the NIT iterations are reached to perform one of the interlacing techniques between the parameters of the threads. For the parallel algorithm presented in this chapter, three methods were established:

- **No restart:** this restarting method runs the algorithm in the number of threads specified and without any kind of reset between the threads.
- **Restart to the best:** in this option, all the threads are restarted with the parameters of the best one.
- **Restart to k bests:** for this restart option, all the threads out of the k_BEST set are restarted to a randomly selected thread of the k_BEST set. The parameter k is the number of threads selected to conform this set.

1.3 Experiments

In this section, 6 benchmark functions were used to determine the best values for the parameters that the Algorithm needs to execute. Once the parameters' values were obtained, the results were compared to the ones published in [5].

Benchmark Functions

As mentioned before, the algorithm was tested with 6 benchmark functions. These functions and their data are shown in Table 1.1.

$$f(x^*) = -d(d+4)(d-1)/6 \tag{1.4}$$

Table 1.1 columns are explained as follows:

- The f column shows the number we gave to the function so that we can refer to them later in the easiest way.
- The $searchspace$ column indicates the interval in which the parameters will be set. If the movement equation makes one parameter go outside the search space interval, we reset it to the closest.
- The D column shows the dimensions for which each function was tested.
- The $Function$ column indicates the name of the function.
- The $Equation$ column shows exactly how the functions were programmed for the experiments.
- Finally, the $f(x)$ column shows the real optimum value of the function.

Table 1.1 Benchmark functions

f	Search space	D	Function	Equation	$f(x)$
1	[−10, 10]	20	Step	$f(x)= \sum_{i=1}^{n} (\lvert x_i + 0.5\rvert)$	0
2	[−2, 2]	20	Sphere	$f(x) = \sum_{i=1}^{n} x_i^2$	0
3	[−10, 10]	20	Sum square	$f(x) = \sum_{i=1}^{n} i x_i^2$	0
4	[−100, 100]	20	Trid10[a]	$f(x) = \sum_{i=1}^{n} (x_i - 1)^2 - \sum_{i=2}^{n} x_i x_{i-1}$	−1520
5	[−0.5, 10]	20	Zakharov	$f(x) = \sum_{i=1}^{n} x_i^2 + (\sum_{i=1}^{n} 0.5 i x_i)^2 + (\sum_{i=1}^{n} 0.5 i x_i)^4$	0
6	[−5, 10]	20	Rosenbrock	$f(x) = \sum_{i=1}^{n-1} [100 (x_{i+1} - x_i^2) + (x_i - 1)^2]$	0

[a] Note: the Trid10 optimum value depends on the dimensions used. This value is calculated as shown in Eq. (1.4)

Experimental Design

Different experiments were made to determine the number of fireflies that will be used and the number of threads.

For the number of fireflies, we ran the algorithm with the *No restart* method with different dimensions and values of N, as shown in Table 1.2. The results obtained were compared to determine the combination of parameters that gave us the greatest amount of best results trying to increase the complexity of the problem.

All the experiments made from now on were executed 10 times. The results presented are the media of the repetitions.

After analyzing the result obtained in Table 1.2, we determined that the best option for the parameters was a 20 D problem with 35 fireflies because it obtained considerably better results with high complexity.

Even with the parameters used by Hashmi Adil Hashmi [5], we obtained so much better results thanks to the modification made in the original algorithm.

The next step for the determination of parameters is to set the number of threads to run the algorithm. This was made similarly to the number of fireflies' experiment. We ran the experiment with different amounts of threads with the restart to best technique and determined which one gave us the best results.

The results of these experiments are reported in Table 1.3. The number of threads that better met the expectations was eight. While increasing the number of threads tested from five to eight, the results improved, when getting the nine-threads results, some of them started to be deficient.

Parallel Firefly Algorithm Final Experiments

As discussed before, once having the analysis of the results obtained in Table 1.3, the best number of threads to run the algorithm was set to **eight**.

The experiment was executed with eight threads and the obtained results are shown in Table 1.4. It is proven that this algorithm generates excellent results compared to other algorithms. Specifically, the results were compared to the ones published by Hashmi Adil Hashmi [5] that tested their algorithm with 10 dimensions and 40 fireflies.

For a better comparison of the results exposed in Table 1.4, the winning algorithm has its result marked with bold characters. As it can be seen in the comparison, the parallel implementation proposed in this chapter obtained better results in four of the six functions tested: Sphere, Sum square, Zakharov, and Rosenbrock.

The parameters used to execute the algorithm are shown in Table 1.5. These were set according to the ones used in different publications.

Table 1.2 Experiments for determining the number of fireflies

	Original FA			Modified FA						
	10 D			10 D			20 D			
f	10 FF	20 FF	40 FF	10 FF	15 FF	20 FF	20 FF	25 FF	30 FF	35 FF
1	9.06E−07	8.046	5.2461	4.12E−06	8.10E−05	3.62E−05	0.0350	0.0340	0.0138	0.0445
2	8.23E−07	1.24E−05	5.70E−06	1.45E−15	1.49E−22	1.69E−29	6.86E−29	7.63E−36	8.39E−43	8.96E−50
3	0.002	9.65E−07	9.37E−07	7.67E−15	7.93E−22	6.95E−29	4.31E−27	4.47E−32	1.28E−36	7.85E−38
4	−209.9	−209.9	−209.9	−210	−210	−210	−678.03	−830.53	−929.03	−994.96
5	−0.09	−0.09	−0.09	2.59E−15	2.56E−22	2.11E−29	3.18E−28	1.47E−34	4.26E−34	2.02E−45
6	−9.97E + 03	−9.97E + 03	−9.99E + 03	1.3262	1.9533	3.7246	18.07	17.87	18.50	18.24

Table 1.3 Results to determine the best number of threads with the restart to best method

	Threads					
f	5	6	7	8	9	10
1	0.0138	0.0024	0.0046	0.0041	0.0033	0.0155
2	6.78E−50	6.72E−50	7.03E−50	6.65E−50	6.39E−50	5.98E−50
3	9.52E−41	3.99E−41	3.93E−41	3.23E−41	5.66E−40	5.15E−42
4	−1045.87	−1022.68	−1076.15	−1015.14	−1038.74	−1021.12
5	1.23E−42	1.57E−46	8.23E−47	4.85E−47	2.02E−42	7.50E−47
6	16.64	16.22	15.98	14.83	16.94	14.65

Table 1.4 Experiments with the restart to *k* bests technique, 8 threads, and 35 fireflies

		Firefly Algorithm	Parallel Firefly Algorithm	
f	Function	Mean	Mean	Sigma
1	Step	**9.37E−07**	0.0018	7.0E−04
2	Sphere	5.2461	**6.88E−50**	3.15E−51
3	Sum square	0.57E−05	**4.13E−41**	1.44E−41
4	Trid 10	−209.9	−1079.11	9.18
5	Zakharov	−0.09	**1.17E−46**	4.68E−47
6	Rosenbrock	−9.99E + 03	**15.11**	0.54

Table 1.5 Parameters used for the experiments

Parameter	Value
Alpha	0.2
Betta	1.0
Gamma	1.0
k_BEST	3
Max_Generations	10,000
NIT	1500

1.4 Conclusions

This new implementation of a Parallel modified Firefly Algorithm has proven to be efficient obtaining significantly better results compared to other published in the literature about Firefly Algorithm. Allowing the last fireflies in the ordered set to have bigger steps than the first ones allows two things:

- The last fireflies have the ability to move with bigger steps trying to reach a better space in the search area of the functions.
- The first fireflies move with little steps to remain inside a potential area and trying to improve the results.

At the beginning of the algorithm, the fireflies move normally to really explore the search space. After half of the iterations are executed, they are already located in potential spaces, so a more refined search is made with the new movement equation presented.

For further work, we seek to develop restart strategies that take advantage of these new modifications and test different moments to start using the new movement equation.

We search to implement this kind of movement that depends on the actual positions of the parameters in other kinds of algorithms because it has proven to generate really good results.

References

1. S. Boyd, N. Parikh, E. Chu, B. Peleato, J. Eckstein, Distributed optimization and statistical learning via the alternating direction method of multipliers. Foundations and Trends in Machine Learning. **3**, 1–122 (2011). https://doi.org/10.1561/2200000016
2. E.F.P. da Luz, J.C. Becceneri, H.F. de Campos Velho, A new multi-particle collision algorithm for optimization in a high performance environment. J. Comput. Interdiscip. Sci. **1**(1), 3–10 (2008)
3. M. Dorigo, C. Blum, Ant colony optimization theory: a survey. Theor. Comput. Sci. **344**(2), 243–278 (2005)
4. R. Eberhart, J. Kennedy, A new optimizer using particle swarm theory, in *Proceedings of the Sixth International Symposium on Micro Machine and Human Science, 1995 (MHS '95)* (1995), pp. 39–43
5. A. Hashmi Adil Hashmi, Firefly algorithm for unconstrained optimization. IOSR J. Comput. Eng. **11**, 75–78 (2013)
6. J.H. Holland, *Adaptation in Natural and Artificial Systems: An Introductory Analysis with Applications to Biology, Control and Artificial Intelligence* (MIT, Cambridge, 1992)
7. H. Huang, H. Qin, Z. Hao, A. Lim, Example-based learning particle swarm optimization for continuous optimization. Inf. Sci. **182**(1), 125–138 (2012)
8. J. JáJá, *An Introduction to Parallel Algorithms* (Addison-Wesley, Reading, 1992)
9. D. Karaboga, B. Basturk, A powerful and efficient algorithm for numerical function optimization: artificial bee colony (abc) algorithm. J. Glob. Optim. **39**(3), 459–471 (2007)
10. S.S.M. Kizhakkethil, Memory based hybrid dragonfly algorithm for numerical optimization problems. Expert Syst. Appl. **83**, 63–78 (2017)
11. J.J. Liang, A.K. Qin, P.N. Suganthan, S. Baskar, Comprehensive learning particle swarm optimizer for global optimization of multimodal functions. IEEE Trans. Evol. Comput. **10**(3), 281–295 (2006)
12. F. Martinez-Rios, A. Murillo-Suarez, A new swarm algorithm for global optimization of multimodal functions over multi-threading architecture hybridized with simulating annealing. Procedia Comput. Sci. **135**, 449–456 (2018). The 3rd International Conference on Computer Science and Computational Intelligence (ICCSCI 2018) : Empowering Smart Technology in Digital Era for a Better Life
13. P.M. Pardalos, J.B. Rosen, *Constrained Global Optimization: Algorithms and Applications* (Springer, Berlin, 1987)
14. C. Sun, X. Yang, T. He, A process planning method based on firefly algorithm. Int. J. Internet Manuf. Serv. **5**, 310 (2018)
15. X.S. Yang, Firefly algorithms for multimodal optimization, in *Stochastic Algorithms: Foundations and Applications*, ed. by O. Watanabe, T. Zeugmann (Springer, Berlin, 2009), pp. 169–178
16. X.S. Yang, X.S. He, *Why the Firefly Algorithm Works?* (Springer, Cham, 2018), pp. 245–259

Chapter 2
A New Approach of the Rain-Fall Optimization Algorithm Using Parallelization

Juan Manuel Guerrero-Valadez and Felix Martinez-Rios

Contents

2.1 Introduction 13
2.2 Rain-Fall Optimization Algorithm 14
2.3 Parallel Implementation of the Rain-Fall Optimization Algorithm 18
2.4 Experiments 21
2.5 Conclusions 22
References 26

2.1 Introduction

This chapter proposes an implementation of nature-inspired optimization algorithms with parallelization. The optimization algorithms find the optimum value of a specific function. Knowing the maximum or minimum value of a specific function is useful in many different fields, like engineering, physics, industrial processes, economics, demography, and computer science. Some application functions are very complex to find their optimum values with the traditional methods. That is why it became a need to develop new algorithms capable of finding those global optimums.

Researchers started proposing different algorithms to find those optimums, but not all of them were as precise as they need them to be. Some algorithms fail to get the correct answer in specific functions. To attend that problem, researchers started to design metaheuristic optimization methods. The term "metaheuristic" refers to general-purpose algorithms applicable to optimization problems, usually without a lot of modifications to adapt to a given specific problem [6]. The algorithm that is implemented for multithreading in this chapter is a nature-

J. M. Guerrero-Valadez (✉)
Universidad Panamericana, Facultad de Ingeniería, Mexico City, Mexico
e-mail: juanmanuel.guerrerovaladez@up.edu.mx

F. Martinez-Rios
Universidad Panamericana, Facultad de Ingenierà, Ciudad de Mèxico, Mèxico
e-mail: fmartin@up.edu.mx

P. Vasant et al. (eds.), *Data Analysis and Optimization for Engineering and Computing Problems*, EAI/Springer Innovations in Communication and Computing,
https://doi.org/10.1007/978-3-030-48149-0_2

inspired optimization algorithm called Rainfall Optimization Algorithm (RFO) written by Kaboli, Sevbaraj, and Rahim. The nature-inspired algorithms imitate the biological, physical, and chemical systems that people can find in nature. This allows researchers to have a large number of inspiration sources [2].

RFO has the advantage of being a fast algorithm in multivariable functions, but tends to fail in some functions; it may throw results that are not as precise as other algorithms because it runs out of iterations to get to the global optimum; or starts moving inside the function in the wrong direction. But the implementation of other techniques in RFO can help to make it more precise. This chapter proposes a parallel execution of the algorithm with two different restarting techniques: restart to the best and genetic restart to the best. The first technique helps with the precision of the algorithm and gives the algorithm more iterations focusing only on the best possible results without sacrificing a lot of computation time by the use of parallel executions. The second technique is also nature-inspired based on the natural choice of the best genes, and this technique helps to make the algorithm more metaheuristic performing better in more functions.

Parallel computers offer a high computational power, which makes them useful for many problems in science and engineering. The objective of this chapter is to present the advantages of RFO with parallelization. The new approach given to RFO intends to use the benefits of parallelization running the algorithm in parallel processes at the same time with fewer iterations on different threads. The results obtained by the threads then are compared to choose the best one or to continue the process with the restarting techniques already mentioned and get more precise results.

The chapter is organized as follows: First, the description and pseudocode of RFO are presented. Next, the parallel implementation and the restart techniques are described. In Sect. 2.3, the performance of the algorithm is evaluated with the benchmark functions shown in Table 2.1 and the results obtained with parallelization are compared with each other and with different parameters, as well as with other algorithms including the single-threaded RFO. Finally, in Sect. 2.5, the conclusions are outlined.

2.2 Rain-Fall Optimization Algorithm

The RFO was inspired by the behavior of the drops. Raindrops naturally trickle down along a slope from a peak, then form the rivers, and always reach to the lowest land points or empty into the sea [6]. This behavior can be implemented in an optimization algorithm, since these drops fall precisely through the relief of nature until they reach a place where they cannot continue to fall, either because they get stuck in a local optimum, such as a lake, or because they were reaching the sea level (the global minimum). The drops always tend to go toward the steepest slopes and this is what RFO simulates.

In order for the drops to reach the sea level, a population of raindrops is generated at the first iteration in random positions of the terrain (the function). Afterward, it

Benchmark Functions

Table 2.1 Benchmark functions

Search space	Dimensions	Name	Test function	Global minimum
[−30, 30]	30	Ackley	$f(x) = -20exp(-0.2\sqrt{\frac{1}{n}\sum_{i=1}^{n} x_i^2}) - exp(\frac{1}{n}\sum_{i=1}^{n} cos2\pi x_i) + 20 + e$	0
[−100, 100]	30	Sphere	$f(x) = \sum_{i=1}^{n} x_i^2$	0
[−4, 4]	4	Kowalik[a]	$f(x) = \sum_{i=1}^{n}\left[a_i - \frac{x_1(b_i^2+b_i x_2)}{b_i^2+b_i x_3+x_4}\right]^2$	0.0003075
[−600, 600]	30	Griewank	$f(x) = \frac{1}{4000}\sum_{i=1}^{n} x_i^2 - \sum_{i=1}^{n} cos(\frac{x_i}{\sqrt{i}}) + 1$	0
[−30, 30]	30	Rosenbrock	$f(x) = \sum_{i=1}^{n-1}\left[100\left(x_{i+1} - x_i^2\right) + (x_i - 1)^2\right]$	0

[a]Note: the Kowalik function depends on a_i *and* b_i *arrays as follows [10]:*
$a = [0, 0.1957, 0.1947, 0.1735, 0.1600, 0.0844, 0.0627, 0.0456, 0.0342, 0.0323, 0.0235, 0.0246]$
$b = [0, 4, 2, 1, 0.5, 0.25, 0.167, 0.125, 0.1, 0.0833, 0.0714, 0.0625]$

is necessary to determine the next position that corresponds to the natural behavior of a raindrop. They must move according to the steepest slope of the radius that surrounds them, to accomplish that, there are several methods. The gradient descent is one of the most common methods, but in this algorithm, another method known as random search is used. For a random search, the algorithm has to generate neighbor points around every drop inside of what could be called a neighborhood. Then the optimization function has to be evaluated in every neighbor point and compared with the original drop to finally move the drop to the lowest position of the neighborhood. This process will continue until the drop reaches the lowest point of the terrain (the global minimum) or gets stuck in a puddle (a local minimum).

Description of the Algorithm To understand Algorithm 1, a description of its components is necessary.

Raindrop A particle in a population that contains the following properties:

X_k: The raindrop position of the *k*th variable of the optimization problem.
Status: The status of the drop (active or inactive) to stop considering the drops that are stuck or far away from the global minimum.
Value: The result of the evaluation of the drop at the current position according to the optimization function.
Rank: The raindrops' rank according to the merit-order list.
Step size: The current step size. As it was mentioned before, the RFO uses random search technique to find the next position of the drop. Depending on the position of the drop, the function may behave differently. To adjust those behaviors, it is necessary to change the size of the neighborhood size [11].

Neighborhood Space where the neighborhood point of the *N*th drop can be generated according to Eq. (2.1) [6]:

$$D_k + (r_{initial} * r_N) \tag{2.1}$$

Where:

- **D_k:** Drop position at *k*th dimension.
- **$r_{initial}$:** The initial neighborhood size according to Eq. (2.2).
- **r_N:** The step size of the drop *N*. Generated according to Eq. (2.3):

$$r_{initial} = (|up_k| + |low_k|) * 0.02 \tag{2.2}$$

$$r_N = r_{last} * (1 + U) \tag{2.3}$$

Where:

- **r_{last}** is the step size of the drop *N* at iteration *i - 1*. At iteration 1, r_{last} is equal to 1.
- **U** is a random number between −1 and 1.

Neighbor Point New points inside the raindrop's neighborhood are generated randomly and are evaluated to find the next dominant position of the drop. Each neighbor point j of drop i is represented as NP_i^j in Eq. (2.4) [6]:

$$\begin{aligned} & \| \left(D^i - NP_j^i\right).\bar{u_k}\| \leq \|r.\bar{u_k}\| \\ & 1 \leq i \leq m \\ & 1 \leq j \leq np \\ & 1 \leq k \leq n \end{aligned} \tag{2.4}$$

Where:

- **r** is the neighborhood size.

Explosion Process The explosion process is carried out when the drop has no dominant neighbor point. This can occur when the drop generates an insufficient amount of neighbor points or when it gets stuck in a local minimum instead of the global minimum [6]. In the original algorithm, the authors suggest to generate more neighbor points, but this chapter recommends a different method.

In the first test, the algorithm was getting neither improvements with the explosion process nor the expected result. Sometimes that was not caused by an insufficient amount of neighbor points, it was more likely caused by having a wrong step size. A dynamic step size for the explosion process was made as described in Eq. (2.5) to improve the results.

$$exp_{step} = \begin{cases} (0.9)\left(10^{-exp+a}\right)\left(ec^{exp_{base}}\right) & a \geq 1 \\ (0.9)\left(10^{-exp_{base}+1}\right)\left(ec^{exp_{base}}\right) & a < 1 \end{cases} \tag{2.5}$$

Where:

- **a**: can be defined by Eq. (2.6).
- **ec**: is the number of explosion processes that have been carried out in the current drop.

$$a = -\log_{10}\left(\frac{|up_k + low_k|.\frac{3}{200}}{0.9}\right) \tag{2.6}$$

Raindrop's Rank Each drop gets a rank according to Eq. (2.7). The rank determines the raindrop's position at iteration t inside the *merit-order list* [6]

$$D_{rank}^i = \begin{cases} C1_t^i = f\left(D_t^i\right) - f\left(D_1^i\right) \\ C1_t^i = f\left(D^i\right) \\ Rank_t^i = w_1 \times order\left(C1_t^i\right) + w_2 \times order\left(C2_t^i\right) \end{cases} \tag{2.7}$$

Where:

- $C1_t^i$ is the rate of change of value at iteration t for raindrop $\mathbf{D^i}$ with respect to its initial value.
- $C2_t^i$ is the current value for raindrop $\mathbf{D^i}$.
- $\mathbf{Order}C1_t^i$ and $\mathbf{Order}C1_t^i$ are sorted in ascending order.
- w_1 and w_2 are the weighting coefficients that have a constant value of 0.5 [6].
- $\mathbf{Rank^i}_t$ returns the rank for raindrop $\mathbf{D^i}$ at iteration t.

Merit-Order List A list containing the ranks of the raindrops in each iteration, in ascending order. The position of the raindrops in the *merit-order list* is proportional to the number of times each drop will iterate inside the explosion process. So, the drops with higher rank will have the opportunity to do more explosion processes before getting inactive. Also, the drop with the worst rank will change its status to inactive.

2.3 Parallel Implementation of the Rain-Fall Optimization Algorithm

Two new methods of parallel computing were implemented in order to improve results using RFO. The method starts with the generation of multiple rainfalls in different threads but with a lower number of maximum iterations per thread to balance CPU time and with the improvements of the results.

Moreover, another iteration counter called H needs to be created. It will count the number of times that the threads have run the RFO. At the first H iteration, as it is described in Algorithm 1, the initial raindrops' positions are going to be randomly generated inside the search space. But if $H > 1$, the start position will not be random. Instead of that, Algorithm 2 is going to take advantage of the results that were obtained at $H - 1$ iteration according to the chosen method. This chapter suggests two methods: Restart to the best and genetic restart to the best.

Restart to the Best

Restart to the best might be one of the simplest approaches. It consists of comparing among all the global minimums of the results of all threads at $H - 1$ to pick the best drop, as it is shown in Algorithm 3. The best drop will be the one with the lower value. After getting the best drop, instead of generating new random positions to the rainfalls in H, all raindrops of all threads are going to start with the position of the best drop of the last H iteration.

The method is effective, though not perfect. Following this metaheuristic approach might not always be the best technique. The problem with this method is that it behaves like a "glutton" algorithm, seeking the best results in each iteration to choose what seems to be the best [15]. But in some circumstances and functions,

Algorithm 1: Rain-Fall Optimization Algorithm Pseudocode

```
Start;
Set N (Number of raindrops);
Set np (Number of neighbor points per raindrop);
Set MaxIt (Maximum number of iterations);
Set eb explosion base;
Set the search space;
Set Ne = 10 (the number of explosions) and i = 1;
Generate N raindrops randomly inside the search space;
for (i < MaxIt) do
    for (j < N) do
        ec = 0 (explosion counter);
        Generate the np and evaluate the function in every np;
        if there is a dominant neighbor point then
            Change the drop's position to the best neighbor position;
        else
            Modify the drops step size;
            for ec < Ne do
                Apply the explosion process;
                if there is a dominant point then
                    end for;
                else
                    ec++;
                end
            end
            ec = 0 (explosion counter);
            Generate the np;
            Evaluate the function in every np;
            if there is no dominant point then
                Change drop status to inactive;
            end
    Create the merit-order list and change the lowest-ranked raindrop status to inactive;
    Set Ne for each drop according to their rank;
    if All drops are inactive then
        end for;
    end
    i++;
Find the drop with the minimum value;
Print the raindrop position and value as the optimum solution;
End;
```

a worse value at H iteration might be closer to the global minimum than the one that was chosen for having the best position; this bad iteration does backstep in the process to reach the goal.

Algorithm 2: RFO Algorithm in Parallel

```
Start;
MaxH. (Maximum number of H iterations);
Number of threads;
H = 1;
Run the RFO as it is shown in Algorithm 1;
for H < Max.H do
    Get the best drop of H iteration and save them inside of the savedDrops array as it is
    shown in Algorithm 3;
    Run the best drop algorithm 3 to determine the best drop;
    if it is using the genetic restart to the best method then
        Run the genetic algorithm to modify the best drop as it is shown in Algorithm 4;
    end
    H ++;
    Run the RFO from Algorithm 1, but setting all drops to start at the best drop's positions;
end
Compare drops values of each thread to find the drop with the best value;
Print the raindrop position and value as the optimum solution;
End;
```

Algorithm 3: Pseudocode to obtain the best drop of H iteration

```
Start;
Get the savedDrops array;
Sort the drops of the savedDrops array in ascending order according to their value;
Return the first drop of the array as best drop;
End;
```

Algorithm 4: Genetic Algorithm Pseudocode

```
Start;
Get the best drop of the savedDrops array;
for H < randomNumber do
    Define randIndex;
    Define randDrop;
    Replace the best drop position at randIndex with the position at the same randIndex
    of the drop randDrop;
    Evaluate the drop according to the function using the new drops positions;
    if the new value is worse than the last value then
        Recover position of the drops before the modification;
    end
end
Return the new modified drop as best drop;
End;
```

Genetic Restart to the Best

Another approach is a genetic restart. Our method uses the crossover of values ("genes") between two objects of the genetic algorithm, which is also nature-inspired [3, 5]. The implementation of the genetic restart method can be described in Algorithm 4. Before every restart, the algorithm mixes the positions of the best drops of each thread inside the best drop of all threads. If the new position is better, the drop inherits those positions.

2.4 Experiments

To evaluate the performance and effectiveness of the algorithm, it was run 50 times with five different functions. The Rosenbrock function [12], the Ackley function [7], the Sphere function [13], the Grienwank function [8], and the Kowalik function [14]. The first four functions are multidimensional and the Kowalik function is four-dimensional. The code was written with CSharp and executed in Visual Studio with a system having the following characteristics:

- Windows 10 64 bits
- Processor AMD FX-8350 at 4.20 GHz
- 16 GB in RAM

Experimental Results

Table 2.2 shows a comparison between RFO and other optimization algorithms. They were run 1000 times for the multidimensional functions and 50 times for the Kowalik function. The parameters used to get those results are the following:

- Number of drops: 10
- Explosion base: 3
- Maximum number of iterations: 4500

Table 2.2 Different optimization algorithms comparison. The bold values are the best values

Function	GA [1]	PSO [1]	GSO [4]	RFO
Sphere	3.1711	**3.6927e−37**	1.9481e−8	1.4607e−6
Rosenbrock	338.56	37.3582	49.8359	**32.4053**
Ackley	0.8678	1.3404e−3	**2.6548e−5**	1.724e−4
Griewank	1.0038	0.2323	**3.0792e−2**	9.86e−2
Kowalik	7.0878e−3	3.8074e−4	3.7713e−4	**3.1553e−4**

After running the rainfall algorithm over 1000 trials, the raindrops of some runs got stuck in local minimums returning these results. The most problematic function was the Ackley function because it has a large number of local minimums. That failure did not occur very often, not more than 14% of the times on average.

Table 2.3 represents the results obtained after running several times the algorithm in parallel using the *restart to the best* method. And Table 2.4 represents the results obtained using the *genetic restart to the best* method. The algorithm ran in 5, 10, and 15 threads to check if having more threads would help to get better quality results. It also ran the multidimensional functions with 10 and 30 variables and with a different number of raindrops per rainfall. That was also intended to determine which configuration was the most optimum for most cases. Tables 2.3 and 2.4 were obtained, when the program was set to restart the threads 50 times before printing the best drop.

After analyzing the results of Tables 2.3 and 2.4, it was confirmed that the more dimensions the function has, the harder it will be for the algorithm to get the global minimum, having better results in the ten-dimensional functions. Having more dimensions also affected the number of iterations the algorithm needed before ending. On average, the algorithm needed a double amount of iterations to find the global minimum with the benchmark functions set to 30 dimensions.

In both cases, restart to the best and genetic restart to the best, in almost every case, the more population of raindrops the rainfalls has, the closer it gets to the global minimum of the function. But having more raindrops results in a considerable CPU time drop.

Finally, Table 2.5 shows a comparison between all the different methods applied to the rainfall algorithm. In general terms, the *genetic restart to the best* was the technique that threw the best results, obtaining the best ones on three functions. And the results of the *restart to the best* method were better than the original algorithm in two functions. In Table 2.5, it is true that the original algorithm finishes the run with fewer iterations. But having in mind that the algorithm was being restarted 50 times per thread with the parallel methods, the iteration counts showed in Table 2.5 were within the expected results. Also, by running the algorithm in different threads at the same time, the CPU time was not much worse, and both parallel algorithms got better results. And *genetic restart to the best* was more time-efficient than *restart to the best*, as it can be seen from the iteration count.

2.5 Conclusions

The purpose of this chapter was to introduce the implementations of parallel methods for the rain-fall optimization algorithm. To ensure that the algorithm suggested is better for seeking an optimum solution of a function, the algorithm was tested in different benchmark functions with very different behaviors to ensure that it can be used in any case scenario [9]. The results with these new techniques

Table 2.3 RFO performance results in parallel with a restart to the best method

Function	Threads	Global minimum 10D					Global minimum 30D					
		Population				Iterations	Population					Iterations
		5	10	15	20		20	25	30	35	40	
ACKLEY	5	7.02e−4	1.16E+00	1.07e−4	8.23e−5	10,335	3.11e−1	5.59e−4	2.65e−4	5.00e−1	4.68e−1	39,918
	10	1.15e−6	5.33e−7	3.97e−7	2.38e−7	33,197	2.24e−3	4.68e−1	4.66e−1	4.26e−4	4.66e−1	124,619
	15	1.08e−6	6.16e−7	3.96e−7	2.64e−7	41,760	2.28e−3	1.48e−4	1.14e−4	9.92e−5	5.11e−5	158,988
GRIEWANK	5	5.69e−1	4.81e−1	3.67e−1	2.14e−1	3379	2.95e−2	5.67e−2	5.13e−2	5.40e−2	1.11e−2	16,873
	10	2.01e−1	6.41e−2	4.93e−2	4.07e−2	4030	1.09e−4	9.12e−5	4.98e−3	3.74e−3	2.89e−5	6018
	15	9.38e−2	7.63e−2	4.93e−2	4.43e−2	5833	3.83e−3	6.81e−5	4.98e−3	4.51e−5	6.19e−3	8803
KOWALIK	5	6.23e−4	3.22e−4	3.08e−4	3.08e−4	119,748	3.08e−4	3.09e−4	3.10e−4	3.08e−4	3.08e−4	115,201
	10	3.10e−4	3.08e−4	3.08e−4	3.08e−4	216,847	3.08e−4	3.08e−4	3.08e−4	3.08e−4	3.08e−4	190,707
	15	3.08e−4	3.08e−4	3.08e−4	3.08e−4	50,217	3.08e−4	3.08e−4	3.08e−4	3.08e−4	3.08e−4	66,541
ROSENBROCK	5	2.379341	1.435481	2.560348	0.474185	2703	28.06112	23.10978	25.87900	23.80238	23.69382	5008
	10	1.540862	0.976108	0.523402	0.369578	4462	23.14647	23.27184	24.90988	22.64277	23.71649	6305
	15	1.885411	0.661299	0.401941	0.610641	6646	20.04868	23.40882	24.06403	20.09753	21.72781	8662
SPHERE	5	9.10e−5	2.23e−5	9.80e−6	4.82e−6	4465	7.81e−5	4.99e−5	3.39e−5	2.71e−5	1.86e−5	17,992
	10	1.02e−4	2.08e−5	9.47e−6	5.07e−6	9062	6.79e−5	4.95e−5	3.07e−5	2.52e−5	1.73e−5	30,880
	15	8.83e−5	2.16e−5	7.40e−6	3.55e−6	6476	8.05e−5	4.43e−5	3.36e−5	2.32e−5	1.88e−5	8775

Table 2.4 RFO performance results in parallel with genetic restart to the best method

		Global minimum 10D					Global minimum 30D					
		Population					Population					
Function	Threads	5	10	15	20	Iterations	20	25	30	35	40	Iterations
ACKLEY	5	3.89e−4	1.70e−4	9.45e−5	2.87e−7	9576	7.50e−1	7.00e−4	3.26e−2	2.93e−4	2.14e−3	48,378
	10	1.02e−6	5.78e−1	5.78e−1	2.72e−7	38,009	4.68e−1	1.19e−4	1.82e−4	3.59e−1	2.93e−4	123,552
	15	3.70e−4	1.15e−4	1.00e−4	7.80e−5	16,482	4.66e−1	5.02e−5	4.66e−1	3.64e−5	9.31e−1	135,872
GRIEWANK	5	1.18e−1	1.43e−1	7.02e−2	5.05e−2	2006	1.65e−4	8.51e−5	5.62e−5	3.75e−3	3.36e−5	4428
	10	1.43e−1	1.83e−1	1.21e−1	3.69e−2	4534	1.46e−4	7.96e−5	6.22e−5	4.97e−3	2.09e−2	8707
	15	3.06e−1	1.45e−1	5.66e−2	6.16e−2	7286	1.20e−4	3.78e−3	5.45e−5	3.99e−5	9.89e−3	11,894
KOWALIK	5	3.08e−4	3.08e−4	3.08e−4	3.08e−4	29,819	3.08e−4	3.08e−4	3.08e−4	3.08e−4	3.08e−4	32,190
	10	3.08e−4	3.08e−4	3.08e−4	3.08e−4	55,686	3.08e−4	3.08e−4	3.08e−4	3.08e−4	3.08e−4	69,963
	15	3.08e−4	3.08e−4	3.08e−4	3.08e−4	67,100	3.08e−4	3.08e−4	3.08e−4	3.08e−4	3.08e−4	88,001
ROSENBROCK	5	4.160369	0.366293	0.620290	0.412899	2777	25.77538	24.78764	25.01837	19.12900	23.51529	4488
	10	2.023165	0.850127	0.473295	0.342847	5164	25.90955	24.29403	21.28876	21.43533	23.32402	7242
	15	1.411468	0.470433	0.355201	0.339875	8412	23.81455	21.88443	24.60840	21.61806	22.79322	10,923
SPHERE	5	1.09e−4	1.94e−5	9.93e−6	4.76e−6	2437	5.62e−5	5.22e−5	3.58e−5	2.52e−5	1.93e−5	3540
	10	9.67e−5	1.69e−5	1.16e−5	4.45e−6	4322	6.51e−5	5.10e−5	3.07e−5	1.95e−5	1.72e−5	6641
	15	1.07e−4	1.65e−5	6.06e−6	3.88e−6	6928	6.52e−5	4.03e−5	3.04e−5	2.27e−5	1.65e−5	12,154

Table 2.5 Comparison of the RFO using the different techniques. The bold values are the best values

Function	Dimensions	Method					
		Single Threaded		Restart to the best		Genetic restart to the best	
		Global minimum	Iterations	Global minimum	Iterations	Global minimum	Iterations
ACKLEY	30	**1.72e−4**	369	1.79e−1	107,842	2.32e−1	102,601
GRIEWANK	30	9.86e−2	376	1.51e−2	10,565	**2.94e−3**	8343
KOWALIK	4	3.16e−4	3457	3.22e−4	124,149	**3.08e−4**	63,384
ROSENBROCK	30	32.41	429	23.44	6658	**23.28**	7551
SPHERE	30	**1.46e−6**	1020	3.99e−5	19,216	3.65e−5	7445

have proven to obtain better results than the original algorithm. In fact, threads help the algorithm to get better optimum values more efficiently because they allow the program to allocate more rainfalls and more raindrops. And as Algorithms 3 and 4 illustrate, having more raindrops increases the chances of getting a better result. And the restart to the best and genetic restart to the best force the algorithm to keep finding better solutions and getting the drops closer to the global minimum after every restart.

References

1. R.C. Eberhart, Y. Shi, Comparison between genetic algorithms and particle swarm optimization, in *Lecture Notes in Computer Science* (Springer, Berlin, 1998), pp. 611–616. https://doi.org/10.1007/bfb0040812
2. I. Fister Jr., X.S. Yang, I. Fister, J. Brest, D. Fister, A brief review of nature-inspired algorithms for optimization. Elektrotehniski Vestnik/Electrotechnical Review. **80** (2013)
3. G. Harik, The compact genetic algorithm. IEEE Trans. Evol. Comput. **3**(4), 287–297 (1999)
4. S. He, Q. Wu, J. Saunders, Group search optimizer: an optimization algorithm inspired by animal searching behavior. IEEE Trans. Evol. Comput. **13**(5), 973–990 (2009). https://doi.org/10.1109/tevc.2009.2011992
5. E. Hou, A genetic algorithm for multiprocessor scheduling. IEEE Trans. Parallel Distrib. Syst. **5**(2), 113–120 (1994)
6. S.H.A. Kaboli, J. Selvaraj, N. Rahim, Rain-fall optimization algorithm: a population based algorithm for solving constrained optimization problems. J. Comput. Sci. **19**, 31–42 (2017)
7. D. Karaboga, B. Basturk, A powerful and efficient algorithm for numerical function optimization: artificial bee colony (ABC) algorithm. J. Glob. Optim. **39**(3), 459–471 (2007). https://doi.org/10.1007/s10898-007-9149-x
8. M. Locatelli, A note on the griewank test function. J. Glob. Optim. **25**(2), 169–174 (2003). https://doi.org/10.1023/a:1021956306041
9. T.N. Malik, A. ul Asar, M.F. Wyne, S. Akhtar, A new hybrid approach for the solution of nonconvex economic dispatch problem with valve-point effects. Electr. Power Syst. Res. **80**(9), 1128–1136 (2010). https://doi.org/10.1016/j.epsr.2010.03.004
10. H. Schichl, A. Neumaier, The NOP-2 modeling language, in *Applied Optimization* (Springer, Berlin, 2004), pp. 279–291
11. Schumer, M.: Adaptive step size random search. IEEE Trans. Autom. Control **13**(3), 270–276 (1968)
12. Y.W. Shang, Y.H. Qiu, A note on the extended rosenbrock function. Evol. Comput. **14**(1), 119–126 (2006). https://doi.org/10.1162/evco.2006.14.1.119
13. K. Tang, X. Li, P.N. Suganthan, Z. Yang, T. Weise, Benchmark functions for the CEC'2008 special session and competition on large scale global optimization. Technical report. University of Science and Technology of China. 1 (2009)
14. D. WINFIELD, Function minimization by interpolation in a data table. IMA J. Appl. Math. **12**(3), 339–347 (1973). https://doi.org/10.1093/imamat/12.3.339
15. C. Zanchettin, Feature subset selection in a methodology for training and improving artificial neural network weights and connections, in *2008 IEEE International Joint Conference on Neural Networks (IEEE World Congress on Computational Intelligence)* (IEEE, New York, 2008), pp. 1951–1958

Chapter 3
An Heuristic Algorithm to Calculate the Minimum Amount of Anesthesia Using MRI

Enrique González-Martín, Guillermo Sosa-Gómez, and Omar Rojas

Contents

3.1 Introduction 27
3.2 Materials and Computational Methodology 30
3.3 Metrics for the Evaluation of Segmentations Obtained Through a Computational Process 34
3.4 Metrics 35
3.5 Evaluation of Results 36
3.6 Conclusions 37
References 37

3.1 Introduction

Recent studies assert that the amount of anesthesia to be used depends on many factors, among which stand out [1, 2]: (a) The relative power of the local anesthetic to be used; (b) a number of nodes or spinal nerve roots to block; and (c) a minimum effective concentration of local anesthesia (CMEA). In this chapter, we present a heuristic algorithm used to determine the minimum concentration of local anesthesia. Such amount can be determined, in a relatively accurate way, through magnetic resonance imaging, using image processing techniques such as image registration and *growing regions*.

Cerebrospinal Fluid (CSF) is found in the spinal cord and the brain. It is a colorless, ultra filtrated plasma, that contains glucose, electrolytes, amino acids and other small molecules. It has very few proteins and cells. CSF protects the central nervous system from damage by collision with the surrounding bone structure [3].

E. González-Martín
Universidad Central de Las Villas, Santa Clara, Cuba

G. Sosa-Gómez · O. Rojas (✉)
Universidad Panamericana, Escuela de Ciencias Económicas y Empresariales, Zapopan, Jalisco, México
e-mail: gsosag@up.edu.mx; orojas@up.edu.mx

P. Vasant et al. (eds.), *Data Analysis and Optimization for Engineering and Computing Problems*, EAI/Springer Innovations in Communication and Computing,
https://doi.org/10.1007/978-3-030-48149-0_3

It is necessary to know the volume of cerebrospinal fluid before giving a patient a local anesthetic because the minimum effective concentration of the local anesthetic depends on the volume of cerebrospinal fluid (CSF) where it will be deposited. If the same dose is applied, in equal volume and concentration of local anesthetic, to two patients with different volumes of cerebrospinal fluid, in which the volume is lower, the concentration of the local anesthetic will be higher and therefore the rostral diffusion of the drug will be greater, reaching a higher level of blockade in relation to the one with a higher volume of CSF [4, 5].

Magnetic resonance imaging is a technique in which the patient is placed in a very strong magnetic field and radiofrequency waves are passed through the patient's body in short pulses. Each pulse causes a response pulse of radiofrequency waves that are emitted by the patient's tissues. The place from which these signals originate and their strength is determined by a computer, which produces a two-dimensional image of a section of the patient. MRIs can be produced in any plane and are produced in the form of a cut, which is no more than a portion of the area from which the image is being taken [6, 7]. See Fig. 3.1, where the possible planes (coronal, sagittal and transverse) obtained from an MRI are shown.

By applying different gradients in the height, width, or depth at different times, a complete three-dimensional map of the region of interest can be determined. Furthermore, the obtained images make it possible to obtain information on intensity, form, and even volume of a given region, and thus turn out to be a good option to calculate the CSF's volume.

In order to calculate the volume of the CSF it is necessary to obtain, in the best possible way, the portion of the image with which you wish to work, in this case, the CSF region, this process is called segmentation, which is nothing more than

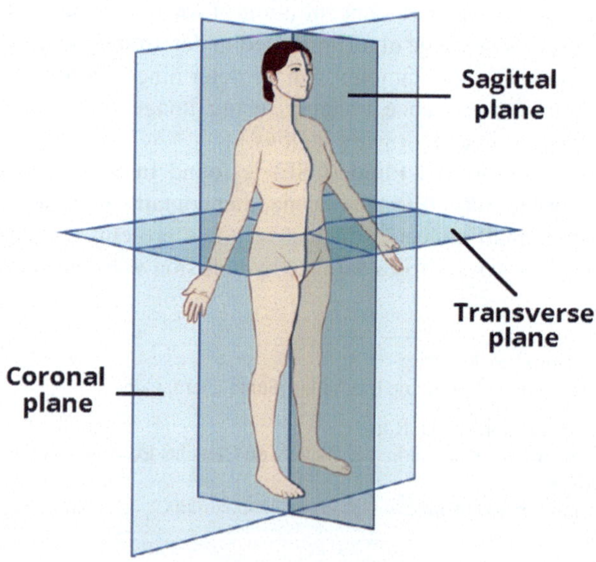

Fig. 3.1 Planes or types of cuts obtained by MRI

the process of dividing a digital image into multiple segments (sets of pixels, also known as superpixels). The goal of segmentation is to simplify and/or change a representation of an image into something that has better meaning and is easier to analyze. Image segmentation is typically used to locate objects and neighborhoods (lines, curves, etc.) in images. It is the process of assigning a label to each pixel in an image such that pixels with the same label share certain visual characteristics.

Types of MRI Images Used

Images of heavy T1 (T1-weighted MRI): Refers to a set of scanning standards that outline differences between various tissues within the body. In the body, scanners with heavy T1 work well to differentiate fat from water, grease appearing lighter, and water darker [9] (see Fig. 3.2).

Images of heavy T2 (T2-weighted MRI): As in T1 weighing scanners, the fat is differentiated from water, but in this case, the grease appears darker and the water lighter. In the case of the CSF study, it will be clearer in the images of heavy T2. These scanners are particularly good at showing edema [9] (see Fig. 3.3).

Registration of Images

In many image-processing applications, it is necessary to form a pixel-by-pixel comparison of two images of the same field obtained by different sensors, or of two images of the same field obtained by the same sensor at different times. For this, it is necessary to spatially register the images and correct the translation changes, the rotation differences, scale differences, and even the perspective differences

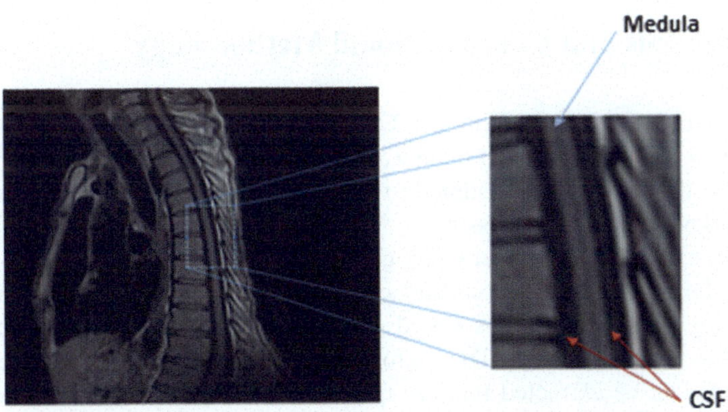

Fig. 3.2 Region of the LCR in MRI type T1

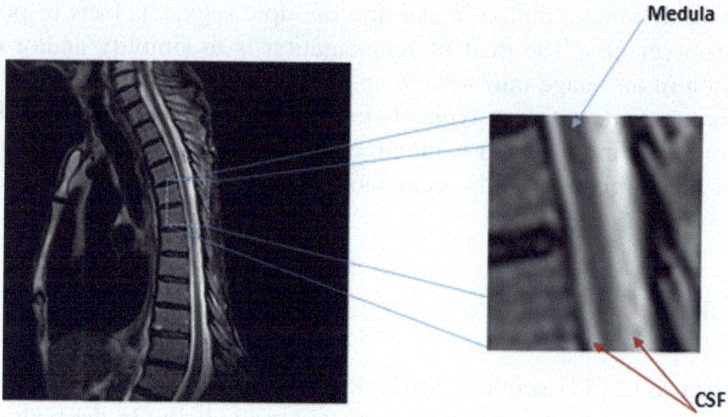

Fig. 3.3 Region of the LCR in MRI type T2

of the images [7]. "Image registration is the process of aligning or developing correspondence between the information of several images" [10].

Connected Component

It is a space that cannot be divided into two sets because each element of that space has some kind of connection with its neighbors; it can also be said that it is a maximal set of nodes such that there is at least one path connecting both nodes. In the case of images, a connected component can be considered as a maximal set of pixels with similar characteristics (in this case intensity) such that there is at least one path that connects both pixels.

3.2 Materials and Computational Methodology

With the objective of segmenting the CSF, the technique presented in this article takes as input two series of MRI images, the images of type T1 and T2, to produce a 3D image with scalar values assigned to the pixels that allow creating a mask using techniques known image processing.

This algorithm uses a pattern recognition heuristic [11, 12]. It was obtained thanks to the results of research and exchanges with experts in medicine, both anesthesiologists and radiologists. CSF is a liquid of uniform density that covers the spinal cord completely. This information is very important since several observations can be extracted that are taken into account during the development of the algorithm:

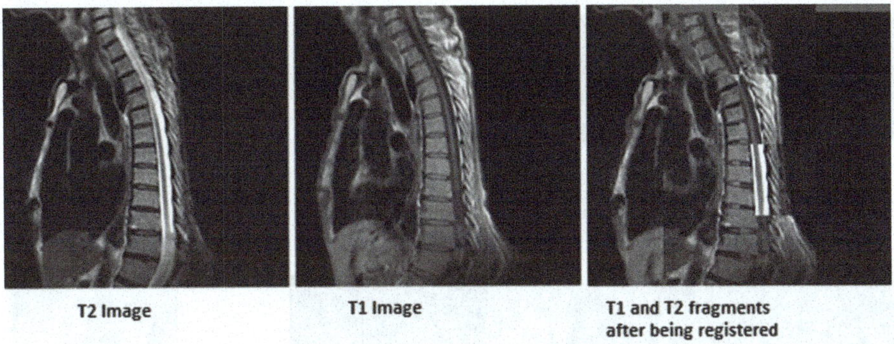

Fig. 3.4 Registration of images

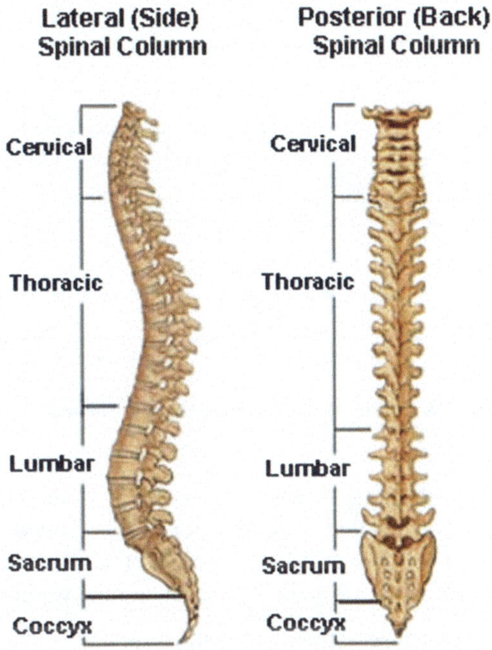

Fig. 3.5 Spine

1. It covers the entire spinal cord, so in the sagittal plane it will look like two thin layers of the same intensity covering a layer of different intensity (see Fig. 3.4).
2. The single and continuous CSF, so it extends in the interior and throughout the extension of the spine (see Fig. 3.5). It is assumed as a related component of uniform density (see Fig. 3.5).
3. When an MRI image is observed from the transverse plane, the marrow is seen as a circular area covered completely by a same thin layer of another intensity, this layer that covers the marrow is the CSF. For example, in T1 it should be seen as a clear point covered by a dark layer (see Fig. 3.6), however, in T2 it looks like a dark spot covered by a layer of clear (Fig. 3.7).

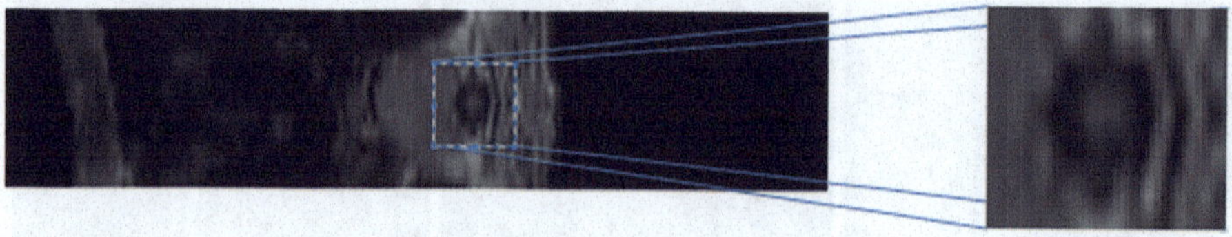

Fig. 3.6 Transverse view of an image T1

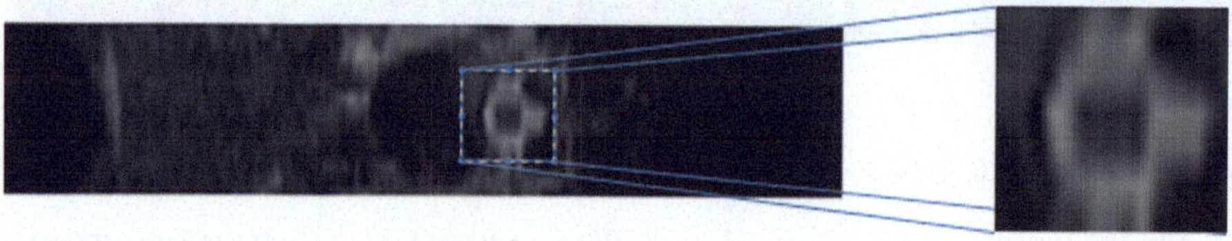

Fig. 3.7 Transverse view of an image T2

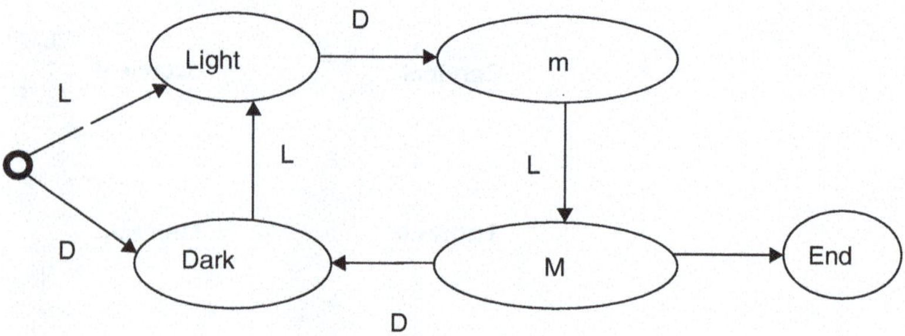

Fig. 3.8 Entries that change the status of the finite state automaton

4. In the images where the column is shown, the marrow is emphasized, so it is expected that the CSF is a related component of considerable length.
5. In the heavy T1, the pixels of CSF are of low intensity and the fat of high intensity, while in T2 the fat and the water have high intensity.

The observations made have served as a guide for the proposed algorithm since the proposed heuristic is based on these to determine the region of the CSF.

Algorithm

To obtain seeds (points where to start looking), the algorithm makes a line scan ("scanline"), executing during the scan the finite state automaton $A(X_{ijk})$. In Fig. 3.8, only the entries that change the state to the PLC are shown, the rest are omitted for clarity.

During the scanline, the automaton is fed with a "L" input whenever a pixel considered "Light" is found (above a certain threshold, i.e. more than 50% of the maximum intensity of the image) and with an input "D" if said pixel is considered "dark" (otherwise):

$$C \ni X_{ijk} : \left\{X_{ijk} > I \max *0, 5\right\}$$

$$O = C'$$

being X_{ijk} is the position pixel i, j, k. Then, if a segment of "scanline" and a given integer value n is found, such that:

$$\left(A\left(X_{ijk}\right) = M\right) \wedge \left(A\left(X_{i,j-n+d,k}\right) = m\right) \forall X_{i,j-n+d,k}; d\,[0..n]\,.$$

That is, a segment that complies with Observation 1: a continuous group of dark pixels surrounded by two continuous segments of clear pixels. So the central pixel of the segment is taken as a seed, formally: $X_{i,\,j\,-\,n/2,\,k}$

This automaton generates multiple false seeds that are discarded taking into account Observation 3. For this, each of the K cross sections is analyzed, generating two regions $R1_K$y $R2_K$ with "region-growing":

$$R1_K = g\left(X_{i,j-n/2,K}, \{O\}\right)$$

$$R2_K = g\left(X_{i,j-n,K}, \{C\}\right).$$

From the above, it can be inferred that $R1_K$ is the possible region of the marrow and $R2_K$ the possible CSF area in section K. Unfortunately, this procedure still generates false positives and regions should be discarded $R1_K$ that is not completely contained in their corresponding $R2_k$. In practice this is done for each K cut, verifying that the minimum dimensions of $R1_K$ are greater than $R2_K$ and its minor maximum dimensions:

$$R1_K^{\max} \leq R2_K^{\max} \wedge R1_K^{\min} \geq R2_K^{\min}$$

$R3$ is defined as a region containing the CSF throughout the volume of the image. You can take as a seed for a "region-growing" of $R3$ any pixel in the set of $R2_K$. If $R3$ contains more than one connected component, the longest one $R3$' is selected.

As stated above, the whole process is done in $T2$ so both the fat and the CSF will be linked. To separate them, the image $T1$ is recorded with $T2$ and the CSF is obtained as the pixels the intersection between $R3'$ and the set of dark pixels of the image $T1$ (O_{T1}):

$$LCR = R3' \cap O_{T1}.$$

Computational Complexity of the Proposed Algorithm

The search and correction of the CSF region have a computational time equal to the dimensions of the image, that is if the image contains 512 × 512 × 14 pixels, then:

Let N, M, O be the number of pixels in each of the 3 dimensions X, Y, and Z respectively, the computational time of the search and correction of the CSF region will be $O\ (N * M * O)$.

The complexity of image registration depends largely on the library used to perform this complex step.

Pseudocode

Read Images T1 and T2
Adjust intensities
Register image T1 with T2
For each sagittal cut in T2 do:
    Until the end of the cut is reached, do:
        Look for the possible CSF area in the sagittal plane (Observation 1)
            If the zone found is completely surrounded (Observation 3)
                Find Related Component
                Rectify the CSF area of the image T1 with T2
Return the largest connected component
Calculate the volume of the component.

3.3 Metrics for the Evaluation of Segmentations Obtained Through a Computational Process

Metrics are used as a measurement criterion to know how good segmentation is. Many times the evaluations are made comparing those obtained in the investigation with existing data and that are assumed as correct. In the case of medical image segmentation, image databases (synthetic or in vivo), segmentation of other techniques, manual segmentation performed by a specialist, etc. are used for comparison. For metallographic images, data obtained from an ISO standard that is known as valid for the materials analyzed can also be used.

The choice of a metric and the validation of an algorithm are not a trivial task. Unfortunately, there are no clear rules as to how to choose an indicator [13]. There are a large number of variables to consider:

- Define the validation method.
- The validity of the data assumed as true.
- Define tolerances at the time of evaluation
- Are all areas of an ROI really equally important?

Given this, the problem can be complicated. In fact, one segmentation can be good enough for a given purpose and not for another. That is why many (or all) of the aforementioned variables are usually defined by the application's users, who have the last word on the decision process. However, it is still necessary to make an estimate of the quality of the segmentation obtained.

To assess the differences between two brain tumor segmentation [14], several similarity coefficients are usually used: Jaccard [15], Hausdorff [16], Dice [17], among others. However, the most common is the Dice Similarity coefficient. This coefficient measures the ratio between the intersection region and the average region:

$$\mathrm{DSC}\ (A, B) = 2\frac{\mid A \cap B \mid}{(\mid A \mid + \mid B \mid)}.$$

For the CSF, this being a relatively new field, no specific comparison forms were found for this anatomical structure.

3.4 Metrics

Manual segmentation by a specialist is still the reference pattern in most cases of medical image segmentation. Unfortunately, it has been seen that it can vary intra- and inter-operator. Therefore, in the evaluation of the Medulla algorithm, it is tried to demonstrate that they behave like one more human operator, verifying that the differences between the algorithm and the human operators do not represent notable differences among themselves. To this end, Friedman's statistical test proposed by [18] is used in his comparative study:

$$\boldsymbol{\chi}_F^2 = \frac{\mathbf{12}\boldsymbol{N}}{\boldsymbol{k}\,(\boldsymbol{k}+\mathbf{1})}\left[\sum_j \boldsymbol{R}_j^2 - \boldsymbol{k}(\boldsymbol{k}+\mathbf{1})^2 * \mathbf{0,\,25}\right],$$

where

$$\boldsymbol{R}_j = \frac{\mathbf{1}}{\boldsymbol{N}}\sum_j \boldsymbol{r}_i^j,$$

r_i^j being the value of the dependent variable j for the ith data set. N is the number of dependent variables. In this case, the data set i is the value of the Dice metric comparing each of the k segmentations performed by the operators (including the algorithm) against that performed by the ith operator. r_i^j is replaced by the value of DSC(i,j):

$$R_j = \frac{1}{N}\sum_j \mathrm{DSC}\,(i,\,j)\,.$$

When interpreting the results, it is considered:

- Highly significant, a significantly less than 0.01
- Significant, a result of significantly less than 0.05 and greater than 0.01
- Mediumly significant, a result less than 0.1 and greater than 0.05
- Not significant, a result greater than 0.1

3.5 Evaluation of Results

To evaluate the marrow segmentation algorithm, we try to confirm that the algorithm behaves like a human operator. Next, the Dice similarity values obtained for the evaluated cases are shown in Table 3.1.

With an asymptotic significance between 0.05 and less than 0.1, the test shows that there are no significant differences. For $N = 6$, the test statistic is $\chi^2 = 6{,}333$, with 2 degrees of freedom and an asymptotic significance of 0.042. Finally, the CSF volumes obtained by segmentation are compared, as shown in Table 3.2.

In this section, only the results are reported, since it was not possible to find another CSF segmentation algorithm with which to compare the efficiency of the same.

Table 3.1 Dice similarity values obtained by the algorithm for the evaluated cases (%)

Cases	Operator 1 vs operator 2	Operator 1 vs algorithm	Operator 2 vs algorithm
1	72,06	68,08	69,07
2	72,03	70,00	71,08
3	77,05	75,08	73,05
4	70,09	72,05	69,00
5	76,08	72,05	68,02
6	70,08	69,08	70,04

Table 3.2 Volumes of LCR obtained by segmentation

Cases	Operator 1 vs operator 2	Operator 1 vs algorithm	Operator 2 vs algorithm
1	10,395 mL	10,305 mL	7,663 mL
2	9,245 mL	13,769 mL	16,506 mL
3	15,656 mL	19,564 mL	20,489 mL
4	11,789 mL	15,369 mL	12,467 mL
5	11,789 mL	15,369 mL	12,467 mL
6	18,276 mL	17,438 mL	16,713 mL

3.6 Conclusions

Despite not being multiple test cases or another similar algorithm with which to compare the proposed algorithm, it has been possible to perform a comparison of the behavior of said algorithm with respect to the response that a human being would provide, and it could be demonstrated that there are no differences significant between the algorithm and a human being. The algorithm can be considered as the first step in order to determine the volume of the LCR efficiently and semi-automatically.

References

1. R.L. Carpenter, Q.H. Hogan, S.S. Liu, Does the variability in the volume of lumbosacral cerebrospinal fluid affect sensory block extent of spinal anesthesia? Anesthesiol.: J. Am. Soc. Anesthesiol. **90**(3), 923–924 (1999)
2. H. Higuchi, Y. Adachi, T. Kazama, The influence of lumbosacral cerebrospinal fluid volume on extent and duration of hyperbaric bupivacaine spinal anesthesia: a comparison between seated and lateral decubitus injection positions. Anesth. Analg. **101**(2), 555–560 (2005)
3. R.A. McPherson, M.R. Pincus, *Henry's Clinical Diagnosis and Management by Laboratory Methods E-Book* (Elsevier Health Sciences, Amsterdam, 2017)
4. D. Longnecker, in *Anesthesiology*, (McGraw-Hill, New York 2008), pp 425–428.
5. M. Galindo-Arias, Seguridad en anestesiología: ¿Qué hay de nuevo? Revista Mexicana de Anestesiología **30**(S1), 71–74 (2007)
6. R.C. Gonzalez, R.E. Woods, *Digital Image Processing* (Prentice-Hall, Upper Saddle River, 2002)
7. W.K. Pratt, *Introduction to Digital Image Processing* (CRC Press, Boca Raton, FL, 2013)
8. W. Allison, *Fundamental Physics for Probing and Imaging* (Oxford University Press, Oxford, UK, 2006)
9. K. Belkic, *Molecular Imaging Through Magnetic Resonance for Clinical Oncology* (Cambridge International Science Publishing, Cambridge, UK, 2004)
10. W. Schroeder, L. Ng, J. Cates, The ITK Software guide second edition updated for ITK version 2.4. FEBS Lett. **525**, 53–58 (2005)
11. J. Kim, J. Pearl, A computational model for causal and diagnostic reasoning in inference systems, in *International Joint Conference on Artificial Intelligence*, ACM, New York, pp. 190–193, 1983
12. G. Gigerenzer, P.M. Todd, *Simple Heuristics That Make us Smart* (Oxford University Press, Oxford, 1999)
13. L. Ibanez, W. Schroeder, L. Ng, J. Cates, *The ITK Software Guide: Updated for ITK Version 2.4* (Kitware, New York, 2005)
14. S.L. Cichosz, S. Vangsgaard, A.S. Jørgensen, K.E. Kannik, E. Steffensen, S.F. Eskildsen, Brain tumor segmentation from MRI: a comparative study, in *IADIS Multi Conference on Computer Science and Information Systems: Computer Graphics, Visualization, Computer Vision and Image Processing, Web Virtual Reality and Three-dimensional Worlds, Visual Communication.* International Association for Development, IADIS, Lisbon, Portugal, pp. 401–406, 2010
15. P. Jaccard, The distribution of the flora in the alpine zone. 1. New Phytol. **11**(2), 37–50 (1912)

16. P. Öberg, Segmentation of the cardiovascular tree in 4D from PC-MRI images. LUTFMA-3244-2013, 2013
17. L.R. Dice, Measures of the amount of ecologic association between species. Ecology ***26***(3), 297–302 (1945)
18. J. Demšar, Statistical comparisons of classifiers over multiple data sets. J. Mach. Learn. Res. **7**, 1–30 (2006)

Part II
Mixed-Integer Programming and Global Optimization

Chapter 4
Backbone Distribution Network Design for the Mexican Automotive Industry

Brenda Retana-Blanco, Jose Antonio Marmolejo-Saucedo, Roman Rodriguez-Aguilar, and Erika Pedraza-Arroyo

Contents

4.1 Background ... 41
4.2 Literature Review ... 42
4.3 Data Analysis ... 44
4.4 Mathematical Model ... 47
4.5 Software Used ... 52
4.6 Location Analysis ... 54
4.7 Results ... 56
References ... 59

4.1 Background

Information from the Database

For this study, the data used were scaled by a factor to not disclose confidential company information. Therefore, the information presented does not correspond to reserved information.

The original version of this chapter was revised: Contributing Authors "Roman Rodriguez-Aguilar, Brenda Retana-Blanco and Erika Pedraza-Arroyo" affiliations have been updated. The correction to this chapter is available at https://doi.org/10.1007/978-3-030-48149-0_21

B. Retana-Blanco · E. Pedraza-Arroyo
Universidad Anáhuac Mexico, Facultad de Ingeniería, Estado de Mexico, Mexico
e-mail: brenda.retana@anahuac.mx; erika.pedrazaar@anahuac.mx

J. A. Marmolejo-Saucedo (✉)
Universidad Panamericana, Facultad de Ingeniería, Ciudad de México, México
e-mail: jmarmolejo@up.edu.mx

R. Rodriguez-Aguilar
Facultad de Ciencias Económicas y Empresariales, Universidad Panamericana, Ciudad de México, México
e-mail: rrodrigueza@up.edu.mx

P. Vasant et al. (eds.), *Data Analysis and Optimization for Engineering and Computing Problems*, EAI/Springer Innovations in Communication and Computing,
https://doi.org/10.1007/978-3-030-48149-0_4

Table 4.1 Sales 2018

Vehicle line	Units 2018 Non-North America (NNA)	Assignment percentage
Ecosport	2,757	10.73
Escape	3,085	12.01
Figo	14,230	55.38
Ranger crew	3,469	13.50
Transit	2,155	8.39
Total	25,696	100

Table 4.2 The database headers

Vehicle key: VIN8
Retail distributor
Catalogue
Wholesale distributor
Vehicle line
Wholesale city
Retail city
State
Payment date
Release date
Date of production
Wholesale invoice date
Retail invoice date

For this project, 50,843.00 data on vehicle demand were collected, which were divided into 3 different origins: Non-North America (Ecosport, Escape, Figo, Ranger, and Transit), North America (Edge, Expedition, Explorer, f-series, Lincoln Continental, Lincoln MKC, Lincoln MKX, Mustang, Navigator, Lobo, and Police Interceptor), and Mexico (Fiesta, Fusion, Fusion Hev, and Lincoln MKZ). This project will only focus on vehicles from Non-North American origin, which reduces the database to 25,696.00 vehicles, with 5 lines of vehicles divided in the following way (see Table 4.1):

For the analysis, the following information (Table 4.2) provided by the company per vehicle unit was considered:

The information provided goes from January 2018 to October 2018, from which the monthly historical demand by distributors was obtained.

4.2 Literature Review

It is a fact that companies are now increasingly concerned with enhancing the efficiency of their processes, as a result of both tactical and operational impacts on the highest-level positions. For instance, once a working structure is selected, all the important decisions within the corporation depend on the framework chosen, and therefore must be directed to fulfill the company's goals. Researchers often include this perspective in the design of the network supply and they usually apply real-life situations to solve the problems that arise. In this context, a document from the former art may be found in [4].

While previously the distance between plants and customers was not considered a key component, new methods take into account the optimal location of the distribution centers as part of the supply chain. It is of particular interest, a study regarding the location of the facilities that explains the design of the backbone distribution network for the automotive industry in Mexico along with some important features, methods, and applications related to the distribution system management [10]. Additionally, authors sustain that the geographical placement of discrete facilities could be classified as network supply design issues.

It is not unusual that several industries experience difficulties when they plan their distribution network since the decisions involved are related to the improvement of the service provided and other important metrics. Among the types of problems commonly encountered in supply chain management, the placement of facilities has been studied for a long time [9]. For example, even though this may vary from one situation to another, many techniques used in the design of the distribution network belong to the research operations field; like the ones presented in [5], whose contributions to this matter may be used as models for the selection of locations. Similarly, a case study from a consumer goods company is presented in [3], where a distribution network model of a region is used to make several precise decisions such as delivery time, credit performance, power, and the reputation of the business. Finally, in [15], the authors aim to propose a new distribution scheme.

The following investigations include representative problems related to the location of the facilities where the case studies exhibited are solved by means of appropriate techniques, according to their characteristics. Typically, these problems are presented as a mixed integer programming formulation (MIP) [7, 13]. In [4], an algorithm based on the decomposition of Benders used to solve the design problem of the distribution network of multiple products is developed. Another classic model is disclosed in [11]. The model shown is used to solve a minimization function that includes the fixed cost in warehouses and distribution centers, as well as the cost of transporting multiple products from plants to warehouses and customers. In addition, the production, distribution, and transportation planning of the multiproduct system is considered as a tri-step scheme. The authors use a heuristic approach based on Lagrangian relaxation to provide an effective and viable solution to the conditions. Moreover, they reconsider other different characteristics to address the main problem of the integrated logistics model [6, 7, 12].

In [14], the case study is settled on the environment of transport planning done by the freight transport industry; hence, an integrated distribution network design and a site selection problem are examined. In this context, a network design of multiple products at a strategic level is contemplated, where each product is defined by a single pair of origin and destination points. Also, the amount of flow required and other implications of the real problem are used in the decomposition of Benders model to find the solution. Furthermore, the authors illustrate the efficiency and effectiveness of this strategy.

Just as in [8], a decomposition of Benders approach is also used in [1]. First, combined with an intelligent algorithm to improve the time solution for the master problem and then, as a modified version to exploit the mathematical formulation of the problem in deterministic, multi-commodity, and single-period contexts, respectively.

Lastly, in [2], the authors stated that these production distribution problems were focused on efficiently finding exact solutions by means of some optimization software (only for small instances and dimensions). However, the main reason to avoid it is it contains a large number of restrictions and variables. Accordingly, they propose a heuristic approach to the two-level mathematical problem using Stackelberg equilibrium.

4.3 Data Analysis

Percentage of Allocation of Sales by City

When performing an analysis of the database, it was obtained that 79% of sales of vehicles was held in 21 cities and the 21% remaining is divided into the other 42 cities (See Table 4.3 and Fig. 4.1).

Assignment of Cases

Based on the concern of the automotive company, the cases were classified as follows:

Success: It is considered success when units were ordered and sold at the same distributor in less than 45 days.

- Error #1: Units that were ordered and sold at the same distributor, but the 45 days credit were overcome.
- Error #2: Units that were ordered and sold at different distributors in the same city, but did the 45 days credit were not exceeded.
- Error #3: Units that were ordered and sold at different distributors in the same city, but the 45 days credit were overcome.
- Error #4: Units that were ordered and sold at different distributors in different cities, but not exceeding from the 45 days of credit.
- Error #5: Units that were ordered and sold at different distributors in different cities, but the 45 days of credit were overcome.

Errors 3, 4, and 5 are considered as the most serious, since they involve overcoming the 45 days of credit and some kind of exchange (see Table 4.4).

Table 4.3 Percentage of allocation by city

Ranking	City	%	Accumulated
1	Mexico City	24%	24%
2	Guadalajara	14%	38%
3	Chihuahua	5%	43%
4	Monterrey	4%	47%
5	Mérida	3%	51%
6	Puebla	3%	54%
7	Canún	3%	57%
8	Hermosillo	2%	59%
9	León	2%	61%
10	Queretaro	2%	63%
11	Saltillo	2%	65%
12	Villahermosa	2%	67%
13	Irapuato	2%	68%
14	Culiacán	2%	70%
15	Cuernavaca	1%	71%
16	Mexicali	1%	73%
17	Cd Obregón	1%	74%
18	Veracruz	1%	75%
19	La Paz	1%	76%
20	Cd. Juárez	1%	78%
21	San Luis Potosí	1%	79%
22	Resto de las ciudades	21%	100%

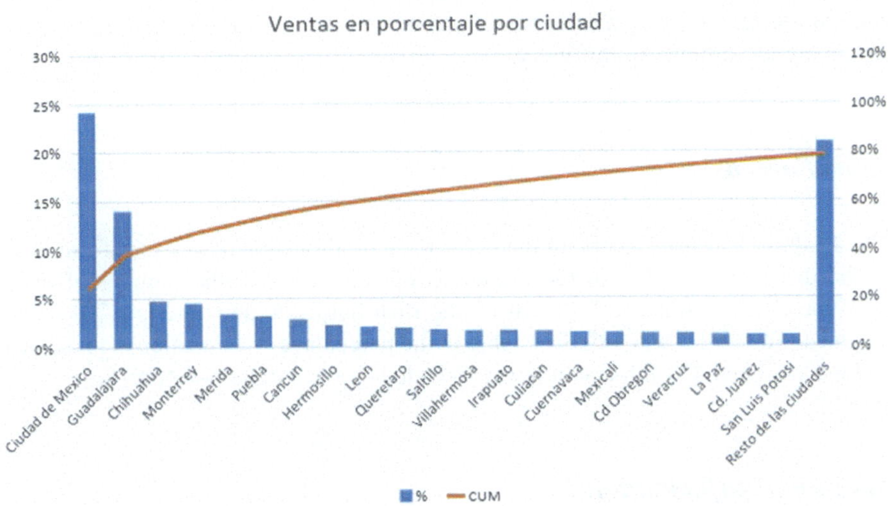

Fig. 4.1 Dealers automotive company

Table 4.4 Success and error table

	$P = V$	$P \neq V$	<45 days	>45 days	Same city	Different city		Percentage
Success	X		X				G	32.37%
Error #1	X			X			Y	37.01%
Error #2		X	X		X		Y	1.94%
Error #3		X		X	X		R	2.43%
Error #4		X	X			X	R	9.82%
Error #5		X		X		X	R	16.43%

From Table 4.4, it can be further enhanced that 32.37% of the units are ordered and sold at the same distributor, without exceeding the 45 days credit. In the same way, it stands out that the most disturbing errors (3, 4, and 5) add up to 28.68%, which could be dramatically reduced if there was a centralized distribution facility, with an optimal location and distribution network. By using the cost information provided by an automotive company and the information obtained in the previous table, the approximate cost that might be caused to the company in 2019 with the same distribution network is calculated. (68,636*26.25%)*$141=$2,516,195.76
Being:

- The number of forecasted sales units in 2019 = 68,636.
- The cumulative percentage of errors 4 and 5 = 26.25%.
- The average transport cost ($) = $141.

By eliminating such errors, the automotive company and its distributors might save approximately USD$2,516,195.76. It is believed that by having a centralized distribution facility, the other types of errors will decrease since distributors will have a higher turnover of inventory.

Sales Forecast

The database was used to obtain the percentage of allocation of vehicles from each of the distributors and with the prognosis of vehicle sales, the number of units by vehicle line estimated for each of the distributors to sell through 2019 was obtained (Tables 4.5 and 4.6). The table in the percentage of allocation is presented in Tables 4.3 and 4.7. Information was provided by the automotive company.

Additional Information

Time in transit: Since the billing and delivery to the distributor dates are available, the average time in days that the automotive company takes to deliver a vehicle in

Table 4.5 Sales forecast 2019

Vehicle line	Sales
Figo	19,597
Ecosport	4181
Escape	3998
Ranger	11,135
Transit	4586
Total	43,497

their different points of sale throughout the country was calculated. The average number of days in transit from wholesale billing to delivery to the distributor was calculated in 8.6 days. This retrieved data will be used as a before and after comparison with the results of the Greenfield Analysis and Network Optimization.

Average sale days: With the analysis performed, it was verified that the average number of days that it takes to sell the vehicle is 101 days, this data meets the information given by the automotive company.

4.4 Mathematical Model

The following analysis aims to define the ideal location for the new production plant, taking into account the place where each of the customers is located and their demands. It is based on the following mathematical method:

$$C_x = \frac{\sum_i d_{ix} w_i}{\sum_i w_i} \quad (4.1)$$

$$C_y = \frac{\sum_i d_{iy} w_i}{\sum_i w_i} \quad (4.2)$$

where:

$d_i x = x$ coordinate of the locality i
$d_i y = y$ coordinate of the locality i
$w_i =$ volume for the locality i

One of the main features of AnyLogistix (ALX) is that it takes into account actual strokes of distances from one point to another. In the same way, it considers suitable transport roads. By using the mathematical method, only coordinates are considered, but not actual distances or roads.

Table 4.6 Sales forecast per vehicle by distributor

Distributor name	Ecosport	Escape	Figo	Ranger	Transit	Total
Productos Automotrices	4	7	116	14	7	148
Country Americas	9	16	97	24	11	157
Automovilistica Andrade	2	3	174	7	1	188
CDA Peninsula	4	5	88	7	1	105
Jimenez Automotriz	3	3	54	1	1	62
CDA Peninsula (Suc. Cancun)	7	4	77	24	1	113
Jalisco Motors	7	10	214	23	5	260
Plasencia Guadalajara	3	3	118	10	7	141
Autos de Hermosillo	7	6	142	11	4	169
Centro Automotriz Vallarta	4	5	195	13	2	219
Sanchez Automotriz	2	3	34	18	2	59
Automotores Coahuilenses	3	5	106	15	1	130
Tabasco Automovilistica	7	8	92	20	2	128
Autos SS de Irapuato	3	5	51	14	1	73
Alden Condesa	5	3	168	12	3	191
Surman Mexico	2	2	64	12	2	82
Autos y Tractores de Culiacan	3	3	42	11	0	59
Car One Valle	0	1	14	3	1	20
Autos y Accesorios	8	5	81	5	114	213
Motores de Morelos	5	5	63	7	0	79
Automotriz del Valle de Baja California	3	2	76	12	3	97
Rivera	5	12	193	46	10	266
Gema Automotores	7	7	165	34	7	220
Zapata	5	4	106	26	3	145
Automotriz Monterrey	4	4	110	26	3	148
Dinastia Automotriz Mexico	4	5	156	43	1	208
Automotriz Baja Cal	4	6	210	14	2	235
Dinastia Automotriz Oaxaca	3	2	127	16	1	150
Acasa Perinorte	5	8	79	8	3	102
Jacobo Rodriguez Motors Piedras Negras	2	7	112	23	22	165
Alden Juárez	4	3	58	29	1	95
Zapata (Suc. Pachuca)	18	7	418	30	4	476
Autos SS de Leon	14	12	347	37	9	418
Automotriz Jalbra	1	2	70	26	4	102
Autos SS de San Luis Potosi	9	5	255	21	6	297
Surman BC	4	3	80	3	2	93
Superservicio	5	3	66	17	3	94
Country Aguascalientes	5	4	103	20	3	136
Gimsa Automotriz	17	17	502	49	6	591
Distribuidores de Autocamiones de Chiapas	4	6	82	9	1	101
Rivera (Suc. Angelópolis)	4	5	127	14	9	159
Vehiculos y Servicios Satelite	7	7	110	24	2	151

(continued)

Table 4.6 (continued)

Leon Automotriz	3	4	64	14	1	87
Automotores Cumbres	7	7	82	26	1	124
Mylsa Queretaro	4	6	98	3	1	112
Cambher Torreon	6	9	138	30	1	184
Montes Queretaro	7	6	118	12	0	144
Autocamiones de Tapachula	5	4	97	2	1	110
Lomas Automotriz	3	5	151	13	1	173
Automotriz Elfer	10	7	367	27	4	414
Camsa	3	5	40	15	1	64
Picacho Automotriz	5	7	449	42	4	507
Ecatepec	5	3	112	17	3	140
Gimsa Automotriz (Suc. Los Reyes)	7	7	87	1	5	107
Plasencia Mazatlan	5	5	86	19	2	117
Cavsa Colima Automotriz	4	4	84	8	0	100
Alden Tlalpan	10	8	129	17	6	170
Cever Durango	3	5	53	10	3	74
Dimotors Coatzacoalcos	5	6	96	15	1	123
Centro Automotriz de la Laguna	2	2	48	8	1	60
Vista Hermosa Laredo Motors	4	8	79	9	3	104
Autos de Calidad de Zacatecas	2	2	51	8	1	64
Motores de Guerrero	9	8	342	28	4	392
Cever Azcapotzalco	4	4	91	12	1	113
Distribuidora Regional	1	2	92	5	0	101
Mylsa (Suc. La Villa)	9	13	526	77	7	632
Automovilistica Andrade (Suc.Tepepan)	1	2	48	14	1	66
Automotriz Monclova	3	4	49	12	1	70
Plasencia Puerto Vallarta	14	5	113	16	7	155
Automotriz Reynosa	4	7	108	13	3	135
Jimenez	9	7	183	27	13	239
Ravisa Motors	3	3	49	8	0	63
Automotriz del Valle de Zamora	5	3	154	10	0	172
Zapata (Suc. Aeropuerto)	2	2	35	10	1	50
Rangel de Alba	4	5	101	24	4	139
Super Autos Jalapa	5	13	147	29	4	198
Automotriz Victoria	8	10	184	46	1	249
Auto Comercial Santa Fe	3	2	43	6	1	55
Plasencia de Nayarit	2	3	25	11	1	43
Mylsa Tehuacan	5	7	106	8	4	131
Vehiculos Automotrices de La Piedad	5	4	84	7	1	100
Vehiculos de Teziutlan	7	11	88	10	12	128
Superservicio Autos del Golfo	3	3	40	9	2	58
Automoviles de Caborca	8	7	78	19	3	114
Total	440	471	10655	1476	378	13420

Table 4.7 Percentage of allocation per vehicle line by distributor

Distributor	Ecosport	Escape	Figo	Ranger crew	Transit	Grand total
Acasa Perinorte	0.11%	0.18%	0.60%	0.13%	0.15%	1.16%
Alden Condesa	0.22%	0.40%	0.51%	0.22%	0.23%	1.58%
Alden Juárez	0.06%	0.07%	0.91%	0.07%	0.02%	1.13%
Alden Tlalpan	0.11%	0.12%	0.46%	0.07%	0.02%	0.77%
Auto Comercial Santa Fe	0.07%	0.07%	0.28%	0.01%	0.02%	0.45%
Autocamiones de Tapachula	0.16%	0.11%	0.40%	0.22%	0.01%	0.91%
Automotores Coahuilenses	0.18%	0.25%	1.11%	0.21%	0.12%	1.86%
Automotores Cumbres	0.07%	0.09%	0.61%	0.09%	0.16%	1.02%
Automotriz Baja Cal	0.17%	0.15%	0.74%	0.11%	0.08%	1.24%
Automotriz del Valle de Baja California	0.11%	0.14%	1.01%	0.12%	0.04%	1.41%
Automotriz del Valle de Zamora	0.05%	0.07%	0.18%	0.17%	0.04%	0.50%
Automotriz Elfer	0.08%	0.12%	0.55%	0.14%	0.01%	0.90%
Automotriz Jalbra	0.17%	0.19%	0.48%	0.18%	0.04%	1.06%
Automotriz Monclova	0.07%	0.12%	0.26%	0.13%	0.02%	0.60%
Automotriz Monterrey	0.12%	0.07%	0.87%	0.11%	0.07%	1.25%
Automotriz Reynosa	0.04%	0.06%	0.33%	0.11%	0.05%	0.59%
Automotriz Victoria	0.07%	0.08%	0.22%	0.10%	0.00%	0.47%
Automóviles de Caborca	0.00%	0.03%	0.07%	0.03%	0.02%	0.16%
Automovilistica Andrade	0.20%	0.14%	0.42%	0.04%	2.53%	3.33%
Automovilistica Andrade (Suc.Tepepan)	0.11%	0.12%	0.33%	0.06%	0.01%	0.63%
Autos de Calidad de Zacatecas	0.07%	0.05%	0.40%	0.11%	0.07%	0.70%
Autos de Hermosillo	0.11%	0.30%	1.00%	0.42%	0.22%	2.06%
Autos SS de Irapuato	0.18%	0.18%	0.86%	0.31%	0.15%	1.67%
Autos SS de León	0.12%	0.11%	0.55%	0.24%	0.07%	1.10%
Autos SS de San Luis Potosí	0.11%	0.11%	0.57%	0.24%	0.07%	1.09%
Autos y Accesorios	0.09%	0.13%	0.81%	0.39%	0.03%	1.44%
Autos y Tractores de Culiacán	0.10%	0.14%	1.09%	0.12%	0.04%	1.50%
Cambher Torreón	0.07%	0.05%	0.66%	0.15%	0.02%	0.96%
Camsa	0.12%	0.21%	0.41%	0.07%	0.06%	0.88%
Car One Valle	0.06%	0.18%	0.58%	0.21%	0.48%	1.51%
Cavsa Colima Automotriz	0.11%	0.07%	0.30%	0.26%	0.02%	0.76%
CDA Peninsula	0.43%	0.18%	2.17%	0.28%	0.08%	3.13%
CDA Peninsula (Suc. Cancun)	0.34%	0.30%	1.80%	0.33%	0.19%	2.97%
Centro Automotriz de la Laguna	0.02%	0.04%	0.36%	0.24%	0.09%	0.75%
Centro Automotriz Vallarta	0.22%	0.13%	1.33%	0.19%	0.13%	2.01%
Cever Azcapotzalco	0.11%	0.09%	0.42%	0.03%	0.04%	0.68%
Cever Durango	0.13%	0.08%	0.34%	0.15%	0.06%	0.76%
Country Aguascalientes	0.13%	0.11%	0.54%	0.18%	0.07%	1.03%
Country Americas	0.42%	0.42%	2.61%	0.45%	0.12%	4.03%
Dimotors Coatzacoalcos	0.10%	0.14%	0.42%	0.08%	0.02%	0.77%
Dinastia Automotriz Mexico	0.09%	0.14%	0.66%	0.13%	0.20%	1.22%
Dinastia Automotriz Oaxaca	0.17%	0.18%	0.57%	0.22%	0.05%	1.20%
Distribuidora Regional	0.08%	0.11%	0.33%	0.13%	0.02%	0.67%

(continued)

Table 4.7 (continued)

Distribuidores de Autocamiones de Chiapas	0.18%	0.19%	0.43%	0.24%	0.02%	1.05%
Ecatepec	0.10%	0.16%	0.51%	0.03%	0.02%	0.81%
Gema Automotores	0.14%	0.23%	0.72%	0.28%	0.02%	1.39%
Gimsa Automotriz	0.17%	0.16%	0.61%	0.11%	0.00%	1.05%
Gimsa Automotriz (Suc. Los Reyes)	0.13%	0.11%	0.51%	0.02%	0.03%	0.79%
Jacobo Rodriguez Motors Piedras Negras	0.06%	0.14%	0.78%	0.12%	0.02%	1.13%
Jalisco Motors	0.23%	0.19%	1.91%	0.25%	0.09%	2.66%
Jimenez	0.07%	0.14%	0.21%	0.14%	0.02%	0.57%
Jimenez Automotriz	0.12%	0.19%	2.33%	0.38%	0.09%	3.11%
Leon Automotriz	0.12%	0.08%	0.58%	0.16%	0.07%	1.01%
Lomas Automotriz	0.18%	0.18%	0.45%	0.01%	0.11%	0.91%
Montes Queretaro	0.12%	0.13%	0.45%	0.17%	0.05%	0.93%
Motores de Guerrero	0.09%	0.10%	0.44%	0.07%	0.00%	0.70%
Motores de Morelos	0.25%	0.20%	0.67%	0.16%	0.12%	1.40%
Mylsa (Suc. La Villa)	0.08%	0.13%	0.28%	0.09%	0.06%	0.63%
Mylsa Querétaro	0.13%	0.15%	0.50%	0.14%	0.02%	0.94%
Mylsa Tehuacan	0.05%	0.05%	0.25%	0.07%	0.02%	0.44%
Picacho Automotriz	0.11%	0.19%	0.41%	0.09%	0.07%	0.88%
Plasencia de Nayarit	0.05%	0.05%	0.26%	0.07%	0.01%	0.45%
Plasencia Guadalajara	0.23%	0.21%	1.78%	0.26%	0.09%	2.56%
Plasencia Mazatlan	0.09%	0.11%	0.47%	0.11%	0.03%	0.81%
Plasencia Puerto Vallarta	0.04%	0.04%	0.48%	0.05%	0.01%	0.61%
Productos Automotrices	0.22%	0.34%	2.73%	0.70%	0.15%	4.15%
Rangel de Alba	0.04%	0.06%	0.25%	0.13%	0.02%	0.49%
Ravisa Motors	0.08%	0.11%	0.26%	0.11%	0.03%	0.58%
Rivera	0.35%	0.12%	0.59%	0.14%	0.16%	1.36%
Rivera (Suc. Angelopolis)	0.11%	0.19%	0.56%	0.12%	0.06%	1.04%
Sanchez Automotriz	0.21%	0.18%	0.95%	0.25%	0.30%	1.89%
Super Autos Jalapa	0.07%	0.08%	0.25%	0.07%	0.00%	0.48%
Superservicio	0.11%	0.08%	0.80%	0.09%	0.00%	1.09%
Superservicio Autos del Golfo	0.04%	0.05%	0.18%	0.09%	0.01%	0.38%
Surman BC	0.11%	0.13%	0.53%	0.22%	0.09%	1.07%
Surman Mexico	0.13%	0.32%	0.76%	0.26%	0.10%	1.57%
Tabasco Automovilistica	0.19%	0.25%	0.96%	0.42%	0.02%	1.84%
Vehiculos Automotrices de La Piedad	0.06%	0.04%	0.23%	0.06%	0.02%	0.40%
Vehiculos de Teziutlan	0.05%	0.08%	0.13%	0.10%	0.02%	0.39%
Vehículos y Servicios Satélite	0.13%	0.18%	0.55%	0.07%	0.09%	1.04%
Vista Hermosa Laredo Motors	0.12%	0.09%	0.44%	0.06%	0.02%	0.73%
Zapata	0.16%	0.29%	0.46%	0.09%	0.27%	1.27%
Zapata (Suc. Aeropuerto)	0.08%	0.09%	0.21%	0.09%	0.04%	0.50%
Zapata (Suc. Pachuca)	0.19%	0.17%	0.40%	0.17%	0.06%	1.00%
Grand total	10.73%	12.01%	55.38%	13.50%	8.39%	100.00%

4.5 Software Used

AnyLogistix (ALX) is a specialized software that serves to address a wide range of supply chain management (SCM) issues, see Fig. 4.2. Decision-making in supply chain management involves the use of quantitative methods, which normally are typically based on optimization or simulation. ALX allows finding appropriate locations and characteristics of the site through the optimization of the network, taking into account demands, capabilities, seasonality, types of products, customer locations, roads, fixed costs, and variable and inventories. The above helps to make an informed decision based on costs, revenue, service levels, utilization rates, and other parameters used in supply chain (Fig. 4.3).

On the other hand, an analysis on how efficient the current distribution network is may be performed and verified if it is reasonable to reset, by opening additional stores or closing any existing. A simulation model of ALX allows capturing the randomness in the behavior of the elements of the chain and enables the evaluation of risks. Also, how the parameters change over time may be assessed; thus identifying cause and effect in network dependencies. ALX combines analytical methods of optimization with simulation in order to have a much deeper analysis. An optimization model describes a network based on the flow of product between locations. In a simulation model, the effect that each variable has within the system may be seen, which gives the possibility to dynamically change parameters during the experiment and to analyze their impact. A simulation does not produce an optimal solution from different combinations, but allows to study in a comprehensive manner particular dynamic scenarios and interdependencies of the supply chain. If more details are considered, then more opportunities to find improvements are created (Figs. 4.4 and 4.5).

Fig. 4.2 AnyLogistix software

Fig. 4.3 Dealers automotive company

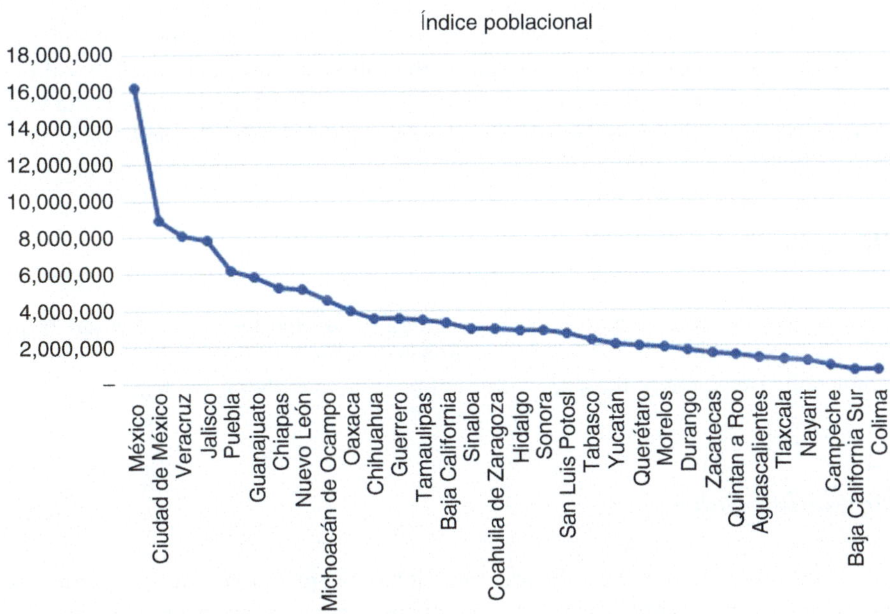

Fig. 4.4 Dealers automotive company

Fig. 4.5 Dealers automotive company

Acomodo de Autos en una Madrina		
Modelo	Foto	Cantidad
Ecosport		9
Escape		8
Figo		11
Ranger		7
Transit		5

4.6 Location Analysis

Several Greenfield Analysis scenarios were performed to obtain the optimal location for the distribution centers, which required the following information. Eighty-three distributors and one supplier were taken into account, which in this case is the Port of Veracruz, since those are Non-North America vehicles (Figs. 4.6 and 4.7).

Historical Data

As enough data were available, it was decided to use the historical demand from January to September, since the information from October did not cover the entire month and also, some of the sales information of the distributors was missing.

Population Index

One of the advantages of the software is that it has an option to perform experiments with population density. Therefore, by conducting a comparison between the population data from INEGI and Figs. 4.1 and 4.4 of vehicle sales by state, it is possible to observe that for greater population densities, there will be higher vehicle sales.

Fig. 4.6 New facility location

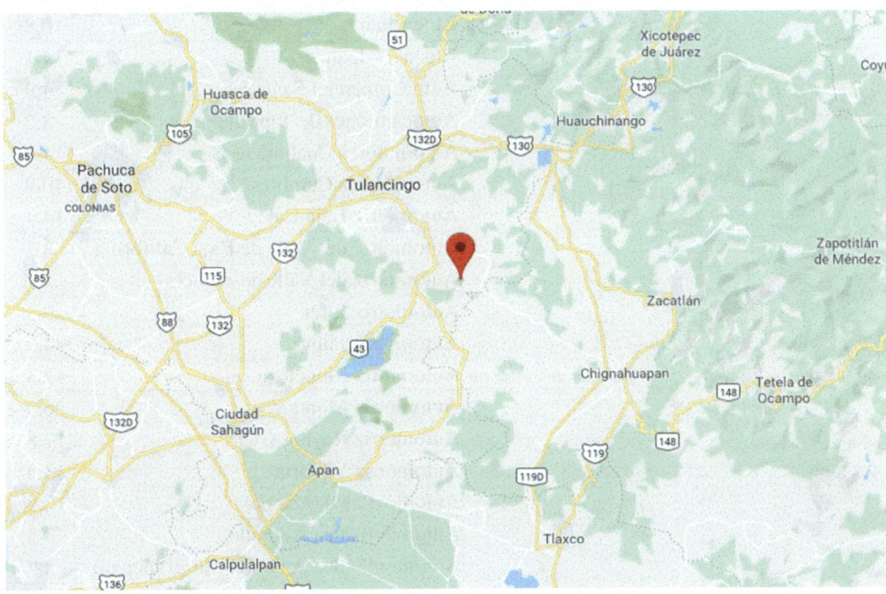

Fig. 4.7 New facility location

Products

As mentioned above, the following lines of vehicles will be used:

4.7 Results

The following results were obtained:

Since the software has geolocation, it is possible to locate the exact coordinates from the distributors.

By using the database above mentioned, the program gave a result of a distribution center at latitude 19.96 and longitude −98.29 coordinates, which correspond to the state of Hidalgo (Table 4.8).

Figure 4.8 shows the distributors, the located distribution center, and the provider that is the Port of Veracruz (Fig. 4.9).

Table 4.8 Average days of delivery

Distributor	Days
Acasa Perinorte	2.8
Alden Condesa	3.5
Alden Juárez	10.8
Alden Tlalpan	4.1
Auto Comercial Santa Fe	4.0
Autocamiones de Tapachula	8.2
Automotores Coahuilenses	14.0
Automotores Cumbres	40.9
Automotriz Baja Cal	14.3
Automotriz del Valle de Baja California	8.3
Automotriz del Valle de Zamora	5.7
Automotriz Elfer	9.3
Automotriz Jalbra	4.9
Automotriz Monclova	12.5
Automotriz Monterrey	26.0
Automotriz Reynosa	17.8
Automotriz Victoria	13.3
Automóviles de Caborca	10.9
Automovilistica Andrade	3.5
Automovilistica Andrade (Suc.Tepepan)	3.9
Autos de Calidad de Zacatecas	13.9
Autos de Hermosillo	8.1
Autos SS de Irapuato	4.2
Autos SS de León	4.0

(continued)

Table 4.8 (continued)

Distributor	Days
Autos SS de San Luis Potosí	8.0
Autos y Accesorios	9.9
Autos y Tractores de Culiacán	9.3
Cambher Torreón	12.7
Camsa	3.0
Car One Valle	11.3
Cavsa Colima Automotriz	9.2
CDA Peninsula	7.1
CDA Peninsula (Suc. Cancun)	8.2
Centro Automotriz de la Laguna	12.8
Centro Automotriz Vallarta	7.3
Cever Azcapotzalco	4.3
Cever Durango	11.7
Country Aguascalientes	10.0
Country Americas	8.6
Dimotors Coatzacoalcos	7.6
Dinastia Automotriz Mexico	3.2
Dinastia Automotriz Oaxaca	4.8
Distribuidora Regional	6.2
Distribuidores de Autocamiones de Chiapas	7.4
Ecatepec	4.0
Gema Automotores	6.4
Gimsa Automotriz	5.0
Gimsa Automotriz (Suc. Los Reyes)	4.4
Jacobo Rodriguez Motors Piedras Negras	12.5
Jalisco Motors	6.7
Jimenez	7.8
Jimenez Automotriz	20.3
Leon Automotriz	6.0
Lomas Automotriz	3.0
Montes Queretaro	5.5
Motores de Guerrero	5.2
Motores de Morelos	3.3
Mylsa (Suc. La Villa)	2.0
Mylsa Querétaro	3.4
Mylsa Tehuacan	5.4
Picacho Automotriz	3.4
Plasencia de Nayarit	9.9
Plasencia Guadalajara	8.4
Plasencia Mazatlan	10.1
Plasencia Puerto Vallarta	11.8
Productos Automotrices	19.0
Rangel de Alba	6.0

(continued)

Table 4.8 (continued)

Distributor	Days
Ravisa Motors	7.4
Rivera	5.2
Rivera (Suc. Angelopolis)	5.0
Sanchez Automotriz	4.1
Super Autos Jalapa	6.9
Superservicio	9.8
Superservicio Autos del Golfo	7.0
Surman BC	10.3
Surman Mexico	4.3
Tabasco Automovilistica	7.6
Vehiculos Automotrices de La Piedad	8.6
Vehiculos de Teziutlan	7.4
Vehículos y Servicios Satélite	2.9
Vista Hermosa Laredo Motors	12.7
Zapata	2.9
Zapata (Suc. Aeropuerto)	4.1
Zapata (Suc. Pachuca)	3.4
Grand Total	8.6

Fig. 4.8 Dealers automotive company

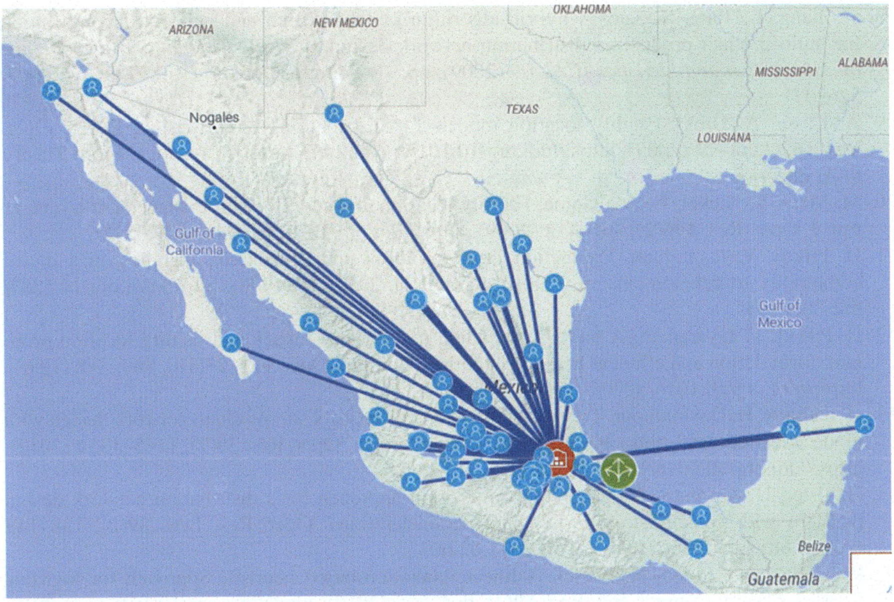

Fig. 4.9 Assignment dealers automotive company

Acknowledgement This project was supported by Universidad Anahuac Mexico and Universidad Panamericana Mexico.

References

1. H.M. Bidhandi, R.M. Yusuff, M.M.H.M. Ahmad, M.R.A. Bakar, Development of a new approach for deterministic supply chain network design. Euro. J. Oper. Res. **198**(1), 121–128 (2009). https://doi.org/10.1016/j.ejor.2008.07.034
2. J.F. Camacho-Vallejo, R. Muñz-Sánchez, J.L. González-Velarde, A heuristic algorithm for a supply chains production-distribution planning. Comput. Oper. Res. **61**, 110–121 (2015). https://doi.org/10.1016/j.cor.2015.03.004
3. A. Cintron, A.R. Ravindran, J.A. Ventura, Multi-criteria mathematical model for designing the distribution network of a consumer goods company. Comput. Ind. Eng. **58**(4), 584–593 (2010). https://doi.org/10.1016/j.cie.2009.12.006
4. A.M. Geoffrion, G.W. Graves, Multicommodity distribution system design by benders decomposition. Manage. Sci. **20**(5), 822–844 (1974). https://doi.org/10.1287/mnsc.20.5.822
5. M.S. Jabalameli, B. Bankian Tabrizi, M. Javadi, Capacitated facility location problem with variable coverage radius in distribution system. Int. J. Indus. Eng. Product. Res. **21**(4), 231–237 (2010)
6. V. Jayaraman, An efficient heuristic procedure for practical-sized capacitated warehouse design and management. Decis. Sci. **29**(3), 729–745 (1998)
7. V. Jayaraman, H. Pirkul, Planning and coordination of production and distribution facilities for multiple commodities. Eur. J. Oper. Res. **133**(2), 394–408 (2001). https://doi.org/10.1016/S0377-2217(00)00033-3. Multiobjective Programming and Goal Programming

8. W. Jiang, L. Tang, S. Xue, A hybrid algorithm of tabu search and benders decomposition for multi-product production distribution network design, in *IEEE International Conference on Automation and Logistics ICAL'09* (2009), pp. 79–84. https://doi.org/10.1109/ICAL.2009.5262971
9. A. Klose, A. Drexl, Facility location models for distribution system design. Eur. J. Oper. Res. **162**(1), 4–29 (2005). https://doi.org/10.1016/j.ejor.2003.10.031. Logistics: From Theory to Application
10. M. Melo, S. Nickel, F.S. da Gama, Facility location and supply chain management a review. Eur. J. Oper. Res. **196**(2), 401–412 (2009). https://doi.org/10.1016/j.ejor.2008.05.007
11. H. Pirkul, V. Jayaraman, Production, transportation, and distribution planning in a multi-commodity tri-echelon system. Transp. Sci. **30**(4), 291–302 (1996). https://doi.org/10.1287/trsc.30.4.291
12. H. Pirkul, V. Jayaraman, A multi-commodity, multi-plant, capacitated facility location problem: formulation and efficient heuristic solution. Comput. Oper. Res. **25**(10), 869–878 (1998). https://doi.org/10.1016/S0305-0548(97)00096-8
13. H. Sadjady, H. Davoudpour, Two-echelon, multi-commodity supply chain network design with mode selection, lead-times and inventory costs. Comput. Oper. Res. **39**(7), 1345–1354 (2012). https://doi.org/10.1016/j.cor.2011.08.003
14. H. Üster, H. Agrahari, A benders decomposition approach for a distribution network design problem with consolidation and capacity considerations. Oper. Res. Lett. **39**(2), 138–143 (2011). https://doi.org/10.1016/j.orl.2011.02.003
15. P.N. Thanh, O. Péton, N. Bostel, A linear relaxation-based heuristic approach for logistics network design. Comput. Ind. Eng. **59**(4), 964–975 (2010). https://doi.org/10.1016/j.cie.2010.09.007

Chapter 5
A Hybrid Model for Improving the Performance of Basketball Lineups

Roman Rodriguez-Aguilar, Rodrigo Infante-Escudero, and Jose Antonio Marmolejo-Saucedo

Contents

5.1 Introduction 61
5.2 Methods 62
5.3 Hybrid Model to Improve the Performance Lineups 65
5.4 Conclusions 70
References 70

5.1 Introduction

The objective of the proposed model is to maximize the performance of a basketball team throughout a game. This is accomplished by answering which lineups must play and how many minutes. The model takes into consideration different factors such as offense, defense, team chemistry, and player fatigue. To achieve this, every lineup is associated with their Net Rating. The Net Rating is a statistic utilized by the NBA that measures a team's point differential per 100 possessions [1].

To run the model for every NBA team, data from the 2016–2017 season, available at stats.nba.com (the official NBA statistics database), was used. It is important to

The original version of this chapter was revised: Editor "Roman Rodriguez-Aguilar" is affiliation has been updated. The correction to this chapter is available at https://doi.org/10.1007/978-3-030-48149-0_21

R. Rodriguez-Aguilar (✉)
Facultad de Ciencias Económicas y Empresariales, Universidad Panamericana, Ciudad de México, México
e-mail: rrodrigueza@up.edu.mx

R. Infante-Escudero
Facultad de Ingeniería, Universidad Anáhuac, Naucalpan de Juárez, Mexico

J. A. Marmolejo-Saucedo
Universidad Panamericana, Facultad de Ingeniería, Ciudad de México, México

P. Vasant et al. (eds.), *Data Analysis and Optimization for Engineering and Computing Problems*, EAI/Springer Innovations in Communication and Computing,
https://doi.org/10.1007/978-3-030-48149-0_5

note that only lineups that participated in at least six games and during a total of a minimum of 35 min were taken into account. This ensures that only meaningful information is used.

Sadiq and Zhao [2] from Deloitte Consulting published a different model to optimize basketball lineups performance titled Money Basketball: Optimizing Basketball Player Selection Using SAS®. This model took a different approach, by obtaining a lineup that maximizes total field goals per game along with court coverage. Player's worth was determined by clustering their shots into "sweet spots" and estimating the number of shots per game within each sweet spot. An optimization algorithm, similar to the maximum coverage problem (MCP), was used to obtain the results. Three different scenarios were explored: one unconstrained, one was a certain player was forced into the solution, and one that considered team's budget. The authors concluded that model only attempts maximizing field goals and does not take into consideration other factors such as rebounding, defense, or free throws. Although, they provided new insight on sports analytics.

Martínez [3] conducted a study that compared different methodologies to assess the player's performance. This article considers individual players indicators rather than those of the team as a whole. Nonetheless, the author assures that traditional box-scores are very limited in terms of how much they actually describe the game. He suggests the use of new methodologies that take a play-by-play approach, such as the adjusted plus-minus.

Brown [4] proposed a new methodology to assess the players' performance. This methodology takes into account different factors such as "smart plays." The model ranks individual players in terms of how much they contributed to successful plays, taking into account the flow of the game. Its limitations include that it can only be applied to single games and that results between games or players are not comparable. The article concludes that the model is a useful alternative to obtain accurate player performance data.

The present work aims to contribute to the generation of the state of the art in the application of mathematical models to sports specifically to basketball. The work is integrated as follows in the first section, the models to be used are described, in the second section, the perspective of the proposed hybrid model is shown, in the fourth section the main results of the application of the model are presented, and finally the conclusions of the study.

5.2 Methods

Optimization Model

A common problem in data analysis is that a large number of variables are handled, some of which may be statistically related (correlated) and, therefore, would be redundant. If n measurements are being made for each of m experimental variables, m vectors (or series of data), each of n components, are generated in m-dimensional space. In this space, each coordinate axis represents a specific variable that is being measured. The redundancy and error (or noise) associated with the measurement process implies that the distribution of data on the m-dimensional space can have a complicated form in which it is difficult to detect useful patterns.

The developed model delivers the optimal use of lineups throughout a 48-min game. The solution reveals how many minutes must each lineup plays in order to maximize net rating. The model's restrictions ensure that the sum of minutes equals 48 and that no player is in the court for more than 34 min. This last constraint is to avoid player's fatigue.

$$\text{Max } z = \sum_{i=1}^{n} k_{ij} m_{ij} \tag{5.1}$$

Subject to:

$$\sum_{i=1}^{n} m_{ij} = 48$$

$$\sum_{p=1}^{n} m_p \leq 34$$

where: z = sum product of minutes given to each lineup and its corresponding net rating; i = lineup number; j = team number; n = total number of lineups; k_{ij} = net rating for lineup i from team j; m_{ij} = minutes for lineup i from team j; m_p = lineups where player p appears.

To obtain the net rating achieved by the selected lineups for a game, the value of z must be divided by 48. A restriction needs to be done for each player, where all the lineups where the player appears are added and this addition requires to be no more than 34.

Technical Efficiency Model: Stochastic Frontiers

Meeusen et al. [5] and Aigner et al. [6] proposed a stochastic model for technical efficiency model. The model includes two components of error technical inefficiency and random error. Forsund et al. [7], Schmidt [8], Bauer [9], and Greene [10] presented the application of stochastic technical efficiency models in different fields. Reifschneider and Stevenson [11] proposed a stochastic technical efficiency model from a production approach applied to a power system in a planning horizon.

Reifschneider and Stevenson [11] and Battese and Coelli [12] proposed a stochastic technical efficiency model where technical inefficiency is explained by a set of endogenous variables and a random error component. Battese and Coelli [13] and Berndt and Wood [14] use a panel data specification according to the following equation.

$$Y_{it} = x_{it}\beta + (V_{it} - U_{it}), i = 1, \ldots, N\ y\ t = 1, \ldots, T \quad (5.2)$$

where: Y_{it} logarithm of the production of the (i) productive unit firm in the time (t); x_{it} vector of inputs for the (i) firm in the time (t); β parameter vector; V_{it} random variable *iid*~N(0, σV^2); U_{it} non-negative random variable (technical inefficiency).

The technical efficiency for the production frontier is defined as:

$$\mathrm{EFF}_i = E\left(Y_i{}^*|U_i, Xi\right)/\mathrm{E}\left(Y_i{}^*|U_i = 0, X_i\right) \quad (5.3)$$

$Y_i{}^*$ is the production of the (i) firm, $\mathrm{EFF}_i \in [0, 1]$ for the production frontier.

Hybrid Model

The proposed model seeks to combine the results of the lineup optimization model and the result of the equipment as an input for the determination of a stochastic frontier. In this case, the optimal lineups derived from the optimization model will be used as the units to be evaluated in the technical efficiency model, as expected; those optimal lineups should be those that show the highest technical efficiency (Fig. 5.1).

It is important to note that the usefulness of this model lies in the monitoring of the performance of the selected optimal lineups, which will allow evaluating if its performance decreases over time, this is possible thanks to the estimation of stochastic boundaries. Additionally, for continuous version of the proposed method, it is possible to use analytical intelligence models to gather prior information on the performance of the lineups as well as the performance during the season.

Fig. 5.1 Proposed model

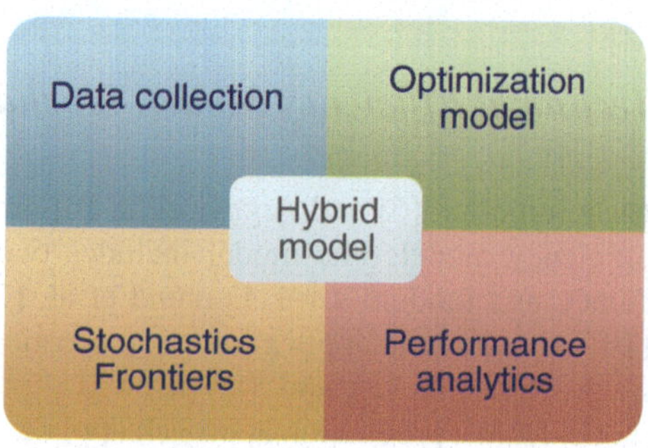

5.3 Hybrid Model to Improve the Performance Lineups

The information used corresponds to official data of the NBA which are available through its website http://stats.nba.com. Information corresponding to the 2016–2017 season was used.

The following criteria were considered for extracting the information of the lineups of each team based on the minutes of play of each player and the number of games played.

- Complete season: minutes $\geq$ 35 and games $\geq$ 6
- Pre-all-star break: minutes $\geq$ 30 and games $\geq$ 5
- Post-all-star break: minutes $\geq$ 16 and games $\geq$ 2

The variables to be taken into account for the optimization model are the offensive score, defensive score, and net score. The analysis was performed for all NBA teams for the 2016–2017 season; there is a direct relationship between the optimal net rating and the net rating of each team (Fig. 5.2).

The outcome variables were considered based on the relationship between net rating and the percentage of matches won (Fig. 5.3).

The optimization model considers the optimal minutes that each lineup should remain and the net rating for that lineup.

A case of Boston is presented to exemplify the application of the proposed model. In the first instance, the optimization model based on the number of games played and the number of minutes per lineup shows us the optimal minutes and in optimal net rating generated by these lineups. This result would suggest that these lineups are the ones that should be more time on the court for their best performance, however, given the physical conditions of the players, this will be a dynamic process between different lineups (Table 5.1).

The results for Boston show that the best lineups or the optimal lineups that should stay longer in the matches are the alienation L8, L14, and L16, Fig. 5.4 shows the detailed performance of these lineups in number of games, minutes played, net rating, and optimal number of minutes per game.

According to the results of the optimization model, if the optimal minutes were aligned, the selected lineups would obtain the following results, where the optimal net rating would be above that observed for the team in the total number of games played (Fig. 5.5).

The example presented for the Boston team shows that the lineups selected as optimal given the net rating observed should be aligned for a greater number of minutes during each match in order to achieve better results in the season. It is necessary to take into account that this model does not consider the possibilities of physical wear of the players, as well as possible injuries, which would modify the feasible alienations in each match.

The estimated stochastic frontiers model considers the net rating as a result variable and as inputs the number of games and the optimal number of minutes,

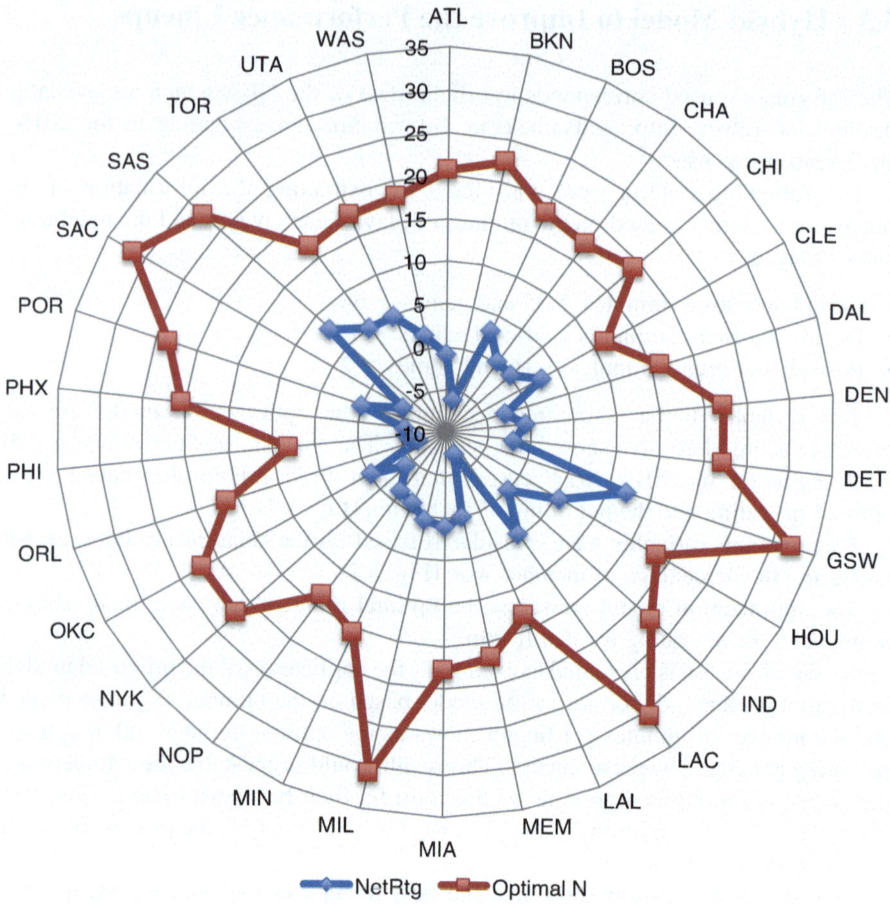

Fig. 5.2 Optimal net rating and net rating

each lineup is considered as a productive unit and the time corresponds to each season recorded for the analysis period.

$$\log\left(\text{Netrating}_{it}\right) = \beta_0 + \log\left(\text{Games}_{it}\right)\beta_1 + \log\left(\text{Optmin}_{it}\right)\beta_2 + \left(V_{it} - U_{it}\right) \tag{5.4}$$

$$i = 1, \ldots, 16 \text{ and } t = 1, 2, 3, 4$$

The results of the optimization model are contrasted with the estimation of efficient frontiers, in this case, the follow-up is given for all the matches of the season and all the alienations. The expected result is that the lineups that are part of the optimization model solution according to the optimal net rating should be those

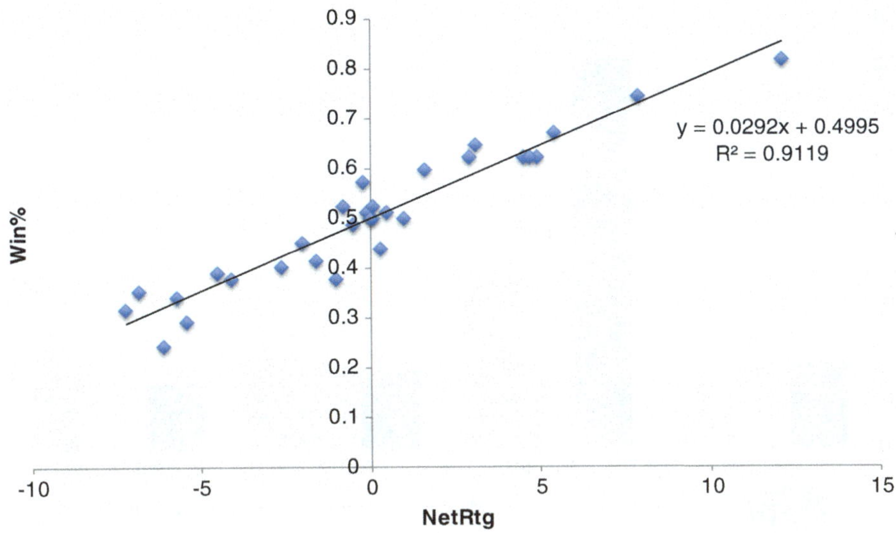

Fig. 5.3 Net rating and games won

Table 5.1 Boston Celtics lineups results

Lineup	PG	SG	SF	PF	C	Games	Min	Net rtg.	Opt min.
L1	I.Thomas	A.Bradley	J.Crowder	A.Johnson	A.Horford	36	433	7.5	0
L2	I.Thomas	M.Smart	J.Crowder	A.Johnson	A.Horford	27	192	16.8	0
L3	I.Thomas	M.Smart	A.Bradley	J.Crowder	A.Horford	32	135	5	0
L4	I.Thomas	M.Smart	J.Crowder	A.Horford	K.Olynyk	25	103	−14.5	0
L5	T.Rozier	M.Smart	J.Brown	J.Jerebko	K.Olynyk	30	99	1.2	0
L6	I.Thomas	J.Brown	J.Crowder	A.Johnson	A.Horford	12	98	12.3	0
L7	M.Smart	A.Bradley	J.Crowder	A.Johnson	A.Horford	16	95	8	0
L8	I.Thomas	A.Bradley	J.Crowder	A.Horford	K.Olynyk	18	82	16.9	14
L9	I.Thomas	M.Smart	A.Bradley	A.Johnson	K.Olynyk	12	82	−12.5	0
L10	I.Thomas	M.Smart	J.Crowder	J.Jerebko	A.Horford	17	56	22.1	0
L11	I.Thomas	A.Bradley	J.Crowder	J.Jerebko	A.Horford	15	53	19	0
L12	I.Thomas	M.Smart	J.Brown	J.Crowder	A.Horford	12	44	0.6	0
L13	T.Rozier	M.Smart	J.Brown	A.Horford	K.Olynyk	14	43	−9.4	0
L14	I.Thomas	M.Smart	J.Brown	J.Jerebko	K.Olynyk	19	42	13.5	14
L15	M.Smart	A.Bradley	J.Crowder	A.Horford	K.Olynyk	16	38	1.5	0
L16	T.Rozier	M.Smart	J.Brown	J.Jerebko	A.Horford	8	38	22.4	20

that have the best performance in the technical efficiency model. The results of the estimation of the stochastic frontier model were showed in Table 5.2.

The stochastic frontier model shows that both the variable number of games and the optimal number of minutes played for each lineup are statistically significant. So, they have a direct implication in the result obtained by each lineup (net rating). The model is significant, and the explanatory variables present the correct sign. The

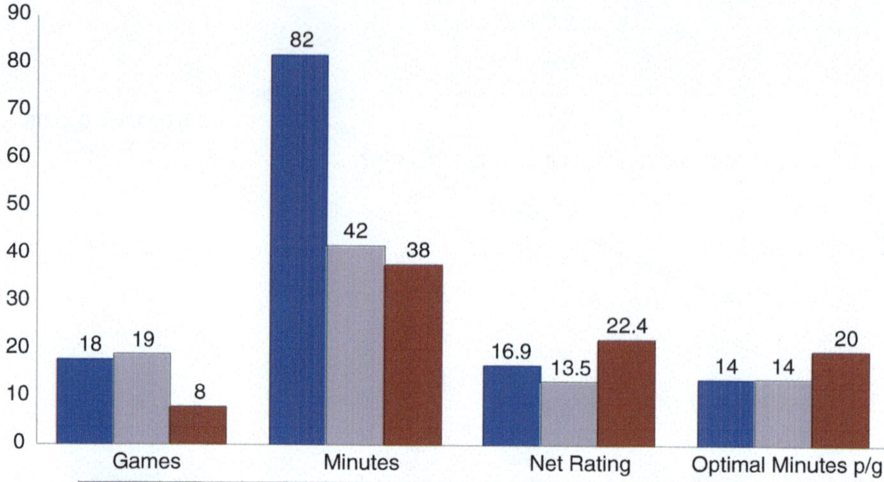

	I8	I.Thomas	A.Bradley	J.Crowder	A.Horford	K.Olynyk
	L14	I.Thomas	M.Smart	J.Brown	J.Jerebko	K.Olynyk
	L16	T.Rozier	M.Smart	J.Brown	J.Jerebko	A.Horford

Fig. 5.4 Lineups comparison for Boston Celtics

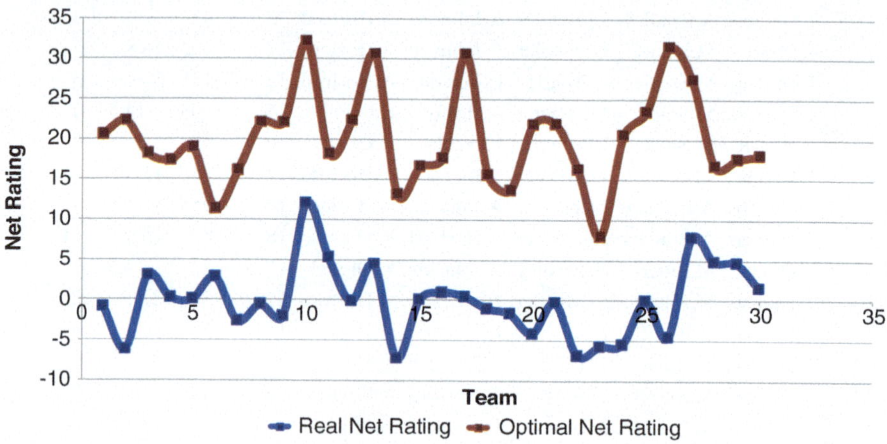

Fig. 5.5 Real net rating vs optimal net rating

estimation of the technical efficiency indicator was performed as invariant for the analysis period; the results of the technical efficiency indicator by lineup are shown in Fig. 5.6.

The results of the technical efficiency estimator based on the estimated stochastic frontier model show that the efficiency of the different lineups is below 50%, it is important to note that the historical data records the lineups used during the

Table 5.2 Stochastic frontier results

Variable	Parameter	S.E	P-value
Games	0.9387209	0.4443667	0.035*
Opt_min	1.557599	0.9772497	0.0111*
Constant	−17.65865	16.22914	0.277
Log-likelihood	−224.1619		
Prob. > Chi2	0.00613*		

The asterisk means that the variable is significant at 95% confidence.

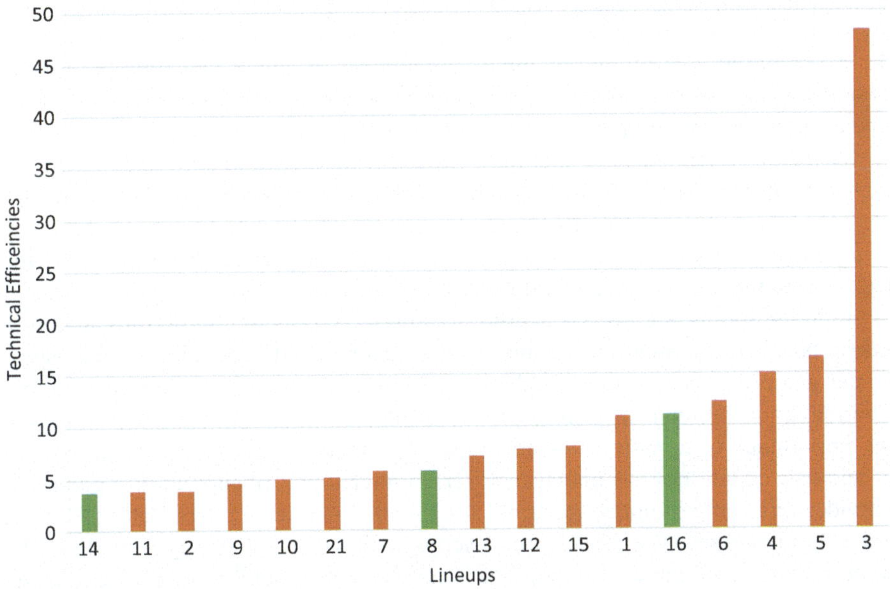

Fig. 5.6 Boston technical efficiencies lineups

matches in the season, In this case, the optimal time per game was considered in the case of the optimal lineups generated by the optimization model. So, it was only possible to consider these optimal times for those parties where the optimal alienation really played. That is why the results show that the lineups with better results in technical efficiency do not correspond directly to the optimal lineups identified in the optimization model. However, it is worth mentioning that lineup 8 and 16 do have technical efficiency above average.

In this case, the model shows its usefulness in the first place to quantify the performance of the different lineups that the team has, in addition to allowing to know the optimal alienations that should play more minutes per game. The estimation of the technical efficiency indicator allows a comparison of the performance of the different line ups. So if you want to evaluate the correct selection of lineups should be selected those that obtained the maximum in the technical efficiency (value of the net rating according to the minutes game optimal).

5.4 Conclusions

This article addressed the application of a hybrid model applied to the investigation of operations in sports. In particular, the performance of different player lineups in the NBA teams for the 2016–2017 season was analyzed. The hybrid model is made up of three axes, (a) the analysis of historical results, (b) an optimization model that maximizes the result of the team based on the lineups used and the minutes played, and (c) a technical efficiency model. The stochastic frontiers model allows monitoring the performance of these lineups by comparing them as productive units between them and ordering them according to their technical efficiency, considering the net rating as a result and as inputs the number of games and time per lineup. The technical efficiency indicator is a way to validate the results of aligning those lineups that were recommended as a result of the optimization model

The results obtained show that the optimal lineups are not necessarily those that were used in the analysis period, so in the technical efficiency analysis these lineups are not in the first places of efficiency as expected. The fact of treating each line up as a productive unit restricts the analysis, since the players can be affected by the lack of rotation and fatigue, in addition to the injuries.

In this proposal, the only net rating was used to measure the performance of a lineup but there are many other metrics that can be used depending on the aspect of the game that wants to be studied as example offensive and defensive rating. The maximum amount of minutes for each player can be adjusted depending on the importance of a game or health status. The model provides a helpful tool for coaches to decide how to manage players' minutes and lineups during games. It provides an average improvement of 20.13 points on a team's Net Rating. Results are only theoretical, does not assure lineups will perform that way in reality. The more minutes and games a lineup plays, its Net Rating will be more dependable. Technical efficiency can be used as another performance measure.

Future work could include finding a more dependable metric to use as coefficients for the objective function. Applying the model to real basketball games would bring lots of insight on how to improve it. Finally, it is necessary to evaluate other methodologies as partial frontier efficiency analysis.

References

1. NBA (NBA Miner, 2018), http://www.nbaminer.com/glossary/net-rating/
2. S. Sadiq, J. Zhao, Money basketball: optimizing basketball player selection using SAS®. (2014), http://support.sas.com/resources/papers/proceedings14/1790-2014.pdf
3. J.A. Martínez, Una revisión de los sistemas de valoración de jugadores de baloncesto (III). Discusión general. Revista Internacional de Derecho y Gestión del Deporte, 12 (2010)
4. S. Brown, A PageRank model for player performance assessment in Basketball, Soccer, and Hockey. (MIT SLOAN, Sports Analytics Conference, 2017), http://www.sloansportsconference.com/wp-content/uploads/2017/02/1494.pdf

5. W. Meeusen, J. Van den Broeck, Efficiency estimation from Cobb-Douglas production functions with composed error. Int. Econ. Rev. **18**, 435–444 (1977)
6. D. Aigner, C. Lovell, P. Schmidt, Formulation and estimation of stochastic frontier production function models. J. Econ. **6**, 21–37 (1997)
7. F. Forsund, C. Lovell, P. Schmidt, A survey of frontier production functions and of their relationship to efficiency measurement. J. Econ. **13**, 5–25 (1980)
8. P. Schmidt, Frontier production functions. Econ. Rev. **4**, 289–328 (1986)
9. P. Bauer, Recent developments in the econometric estimation of frontiers. J. Econ. **46**, 36–56 (1990)
10. W. Greene, The econometric approach to efficiency analysis, in *The Measurement of Productive Efficiency*, ed. by H. Fried, C. Lovell, S. Schmidt, (Oxford University Press, New York, 1994)
11. D. Reifschneider, R. Stevenson, Systematic departures from the frontier: a framework for the analysis of firm inefficiency. Int. Econ. Rev. **32**(3), 715–723 (1991)
12. G. Battese, T. Coelli, *A Stochastic Frontier Production Function Incorporating a Model for Technical Inefficiency Effects. Working Papers in Econometrics and Applied Statistics*, vol 69 (Department of Econometrics, University of New England, Armidale, 1995), pp. 1–22
13. G. Battese, T. Coelli, A model for technical inefficiency effects in a stochastic frontier production function for panel data. Empir. Econ. **20**, 325–332 (1995)
14. E.R. Berndt, D.O. Wood, Technology, prices and the derived demand for energy. Econ. Statist. **57**, 259–268 (1975)

Chapter 6
Location of Bases for Prehospital Services

Zaida E. Alarcón-Bernal, Ricardo Aceves-García, and Jorge L. Rojas-Arce

Contents

6.1 Introduction 73
6.2 Problem Formulation 76
6.3 Solution Method 78
6.4 Case Study 79
6.5 Conclusions 80
References 83

6.1 Introduction

A facility location problem involves many choices at different levels, like deciding the amount and placement of new facilities, distribution of demand nodes to the facilities situated, and planning a transportation network so the most objectives of the system are met. These objectives embrace minimizing total travel time, operational costs, or maximizing market coverage. One application for facility location problems includes the placement of emergency medical services.

In healthcare, the implications of poor location selections transcend price and client service concerns. If few facilities are used or not well-placed, increases in mortality and morbidity will occur. Therefore, facility location modeling becomes even more important when applied to the location of healthcare facilities [7].

Emergency medical services organizations (EMS) are vital parts of health systems, as they are accountable for the prehospital services of health systems, which consist of medical aid and transportation activities performed from the arrival of an emergency call to a patient's release or transfer to a hospital [6]. Therefore, the ability of an EMS to respond efficiently to emergency calls will have a major impact on the health, recovery, and even survival of patients.

Z. E. Alarcón-Bernal (✉) · R. Aceves-García · J. L. Rojas-Arce
National Autonomous University of Mexico, Mexico City, Mexico
e-mail: zaida.alarcon@unam.mx; aceves@unam.mx; jorge.rojas.arce@comunidad.unam.mx

P. Vasant et al. (eds.), *Data Analysis and Optimization for Engineering and Computing Problems*, EAI/Springer Innovations in Communication and Computing,
https://doi.org/10.1007/978-3-030-48149-0_6

Road traffic injuries are a public health problem in Mexico. The economic and social costs we have to pay are very high, so it is necessary to create an emergency response system that categorizes the resources available to serve them. This categorization implies rapid and adequate transportation, as well as medical care on-site and during transportation. However, these systems are expensive, so alternatives must be sought to minimize costs and meet the requested standards.

The emergency service system for responding to traffic accidents is a very important element of an EMS. On the one hand, it is necessary to produce a rapid service to scale back the death rate and so, it is desirable to have adequate capability to deploy the service. This could represent high costs, thus a balance has to be effected between these two components.

This chapter addresses the problem of locating ambulance bases and determining the number of emergency vehicles in order to respond effectively to traffic accidents. Due to the nature of the accidents, demands cannot be known precisely, so possible service demands are considered uncertain. Equity is considered a measure of coverage, as proposed in [1].

The events leading to the intervention of an emergency vehicle at the scene of an incident includes the following steps: (1) incident detection and reporting, (2) call detection, (3) vehicle deployment, and (4) actual paramedic intervention. Decisions made in EMS are related to step two and step three. During the call selection process, the severity of the incident and its degree of urgency must be determined in order to make decision on the type of vehicle to be sent [5]. The U.S. Emergency Medical Services Act of 1973 sets some standards: in urban areas, 95% of requests must be answered within 10 min of the call; in rural areas, they must be answered within 30 min. It is, therefore, necessary to model the location and transfer of emergency vehicles. In addition, considering that uncertainty is a very important component of this problem.

Considering that urban areas are often congested and that response time is a determining factor for the safety and survival of patients and that many of the emergencies can be attended at the site of the accident, this paper will consider the use of traditional emergency vehicles (slower moving ambulances, with at least two paramedics) and motor ambulances (faster moving vehicles, with one paramedic and less equipment).

In recent decades, research into the location of emergency medical services has been extensive. In [2, 5, 13], literature reviews in the field of emergency medical services location are presented. According to the level at which decisions are made, the main models developed are presented in [6]. Some specific proposals can be found at [4, 9, 14, 23].

As a bilevel model, in [11], they propose a location model within which the leader chooses from a series of potential locations to be able to install a service. They consider one or more followers who select their location points once the leader's decision has been made. They present some computational results comparing the formulations and Benders decomposition.

Healthcare systems require efficient use of resources based on clear policies that can be followed. Some of these policies are determined at the strategic level. In [16], a framework for EMS policy making and the use of real-time data analysis and alerts to support decision-making is proposed.

Considering the randomness of demand, in [6], the main works are shown where probabilistic and stochastic models for health services are proposed. Considering stochastic demands, in [15], it is proposed to use sampling to make the model effective. Its proposal is to consider samples of scenarios. In the end, a sample of solutions and a sample of values are obtained.

In Emergency Medical Systems, operators install emergency vehicles in a set of strategic locations to respond to emergency calls, with the intention of minimizing service requests with slow response times. In [3], they propose stochastic formulations for the problem of ambulance deployment. In the proposal, they use emergency call data to model uncertainty. In [8], they propose a two-stage stochastic programming model for relocating and deploying ambulances to maximize expected coverage. Their model restricts staff workload and incorporates call priority levels.

Considering the cooperation between ground ambulances and helicopters in emergency medical rescues in [22], they propose a two-stage coverage location model. First, a coverage model was developed to achieve maximum coverage and minimum total cost of installing emergency rescue facilities. Then, for certain emergency sites, an emergency scheduling mode matrix was constructed to meet the limitations of response time and total rescue time.

Specifically considering traffic accidents, a stochastic scheduling model is proposed in [12] to optimize the operation of an emergency system by adjusting the scheduling plan to locate ambulances when any of them are temporarily unavailable.

In this chapter, a localization model is proposed for dealing with traffic accidents in an urban area, considering that two types of vehicles can be used and that the demand is uncertain. Therefore, a stochastic programming model is proposed, which is resolved by means of the Benders decomposition method, with the contributions:

- We propose a formulation for ambulance bases location and their deployment, that is, a stochastic two-stage location model.
- We used accident incidence data in Mexico City to construct possible scenarios and incorporate uncertainty.
- The model is an extension of the traditional models of location and coverage, which also considers the use of two types of EMS vehicles.

The rest of the chapter is organized as follows: in Sect. 6.2, the description of the problem and its formulation as a two-stage stochastic programming problem is presented. In Sect. 6.3, a solution method for the problem based on the Benders decomposition method is developed, and the results of the computational tests are shown. At the end of this chapter, some conclusions are presented.

6.2 Problem Formulation

The emergency service system for responding to traffic accidents is an important component of an EMS:

- It is necessary for the system to provide a rapid service to reduce the mortality rate, and therefore it is desirable to have sufficient capacity to deploy the service wherever an incident occurs.
- It is necessary to balance these two aspects.

We address the problem of choosing the number of emergency vehicles and the location of their bases in order to have an effective response to traffic accidents. Due to the nature of the accidents, demands cannot be known precisely, so potential service demands are considered uncertain.

In this section, we present the notation and formulation of the facilities location model for the design of emergency services networks for the attention of accidents.

Notation

The sets, parameters, and decision variables used in the proposed model are as follows:

Sets

I Set of eligible locations for emergency vehicle bases
J Set of demand nodes
K Set of emergency vehicle types that can handle the accident

Parameters

f_i Fixed cost of installing a base in the location $i \in I$
a_{ik} Fixed cost associated with each emergency vehicle $k \in K$ at location $i \in I$
c_k Average cost of care per patient per emergency vehicle of type $k \in K$
λ Minimum coverage required
m Maximum number of bases that can be installed
t_{ijk} Travel time of vehicle $k \in K$ from base $i \in I$ to accident point $j \in J$
T Maximum time allowed for arrival to an accident

The last two parameters can be used to define the next set:

$$I_{jk} \quad = \{i \in I : t_{ijk} \leq T\}$$

Since the demand at each time unit is assumed to be stochastic, it will be denoted by s_j, the random variable representing the number of accidents on the site j. A complete realization of the demands, i.e., the number of accidents per point of demand, is a s scenario, and the number of scenarios is finite.

Since the probability distribution of demand at every point is discrete, and the corresponding variable has finite support, we know which situations can occur moreover as the corresponding probability [15]. Then, we can define:

S Set of demand scenarios
d_j^s Number of accidents on the site j on the scenario s
p^s Probability of scenario occurrence s

Each s scenario determines all ambulance requests and the type of requests completely, so when this information is available, we can decide how to assign vehicles.

The independent decisions of the scenarios are those related to the location of the bases and the assignment of vehicles to them, for which the first stage decision variables are:

$$x_i = \begin{cases} 1 & \text{if a vehicle base is located in } i \in I_{jk} \\ 0 & \text{else} \end{cases}$$

w_{ik} : Number of k type vehicles to be installed on site $i \in I_{jk}$

Decisions about deploying vehicles to places where an accident has occurred depend on the scenario that happens, so the second stage decision variables are:

y_{ijk}^s = Proportion of demand of node $j \in J$ which will be attended by a vehicle of type k from base $i \in I_{jk}$ under the scenario $s \in S$.

With these considerations, the problem can be formulated as:

$$minimize \quad \sum_{i \in I_{jk}} f_i x_i + \sum_{i \in I_{jk}} \sum_{k \in K} a_{ik} w_{ik} + \sum_{s \in S} \sum_{i \in I_{jk}} \sum_{j \in J} \sum_{k \in K} p^s (d_j^s c_k y_{ijk}^s) \tag{6.1}$$

$$subject\ to: \quad \sum_{i \in I_{jk}} x_i \leq m \tag{6.2}$$

$$y_{ijk}^s \leq x_i \quad \forall\, i \in I_{jk},\ \forall\, j \in J,\ \forall\, k \in K,\ \forall\, s \in S \tag{6.3}$$

$$\sum_{j \in J} d_j^s y_{ijk}^s \leq w_{ik} \quad \forall\, i \in I_{jk},\ \forall\, k \in K,\ \forall\, s \in S \tag{6.4}$$

$$\sum_{i \in I_{jk}} \sum_{k \in K} y_{ijk}^s \geq \lambda \quad \forall\, j \in J,\ \forall\, s \in S \tag{6.5}$$

$$\sum_{k \in K} w_{ik} \geq x_i \quad \forall\, i \in I,\ \forall\, k \in K \tag{6.6}$$

$$x_i \in \{0, 1\} \quad i \in I_{ik} \tag{6.7}$$

$$w_{ik} \in \mathbb{Z} \quad \forall\, i \in I_{jk},\ k \in K \tag{6.8}$$

$$y_{ijk}^s \in [0, 1] \quad \forall\, i \in I_{jk},\ j \in J,\ k \in K,\ \forall\, s \in S \tag{6.9}$$

The objective function seeks to minimize the fixed costs of installation of the bases and placement of emergency vehicles on them, as well as minimize the costs of accident care. The constraint 6.2 limits the number of bases that can be opened. Constraints 6.3 restrict service only to bases that are open. With 6.4, we limit the number of services available to vehicles by type, both sets of restrictions must be satisfied for all scenarios. The constraints 6.5 force the solution to meet a minimum level of coverage. The restrictions 6.7 force at least one vehicle on each established basis. The restrictions 6.7 limit variables x_i to be binary, 6.8 limit variables w_{ik} to be integer positive and 6.9 to variables y^s_{ijk} to take values between zero and one, since it is the proportion of the demand that will be met.

6.3 Solution Method

Given the formulation of the equivalent deterministic problem, once whole decisions are determined in the first stage, the following stage problem contains solely continuous decision variables. Given the characteristics of the problem, it is appropriate to use the Benders decomposition method [10]. The method iteratively solves the relaxed master problem (RMP) consisting of the first problem and a set of optimality cuts (and/or feasibility cuts) to obtain the first stage solution. Then, the first stage decision solution is used as input into the second stage subproblems, which are solved to provide the RMP cuts.

The main idea of the Benders decomposition method in stochastic programming problems is to approximate the term resource function (second stage cost function) to avoid numerous evaluations of this function. Therefore, we use that term to construct a master problem for first-stage variables, but we only evaluate the resource function as a subproblem. The pseudocode of the algorithm is shown below:

Algorithm 1 Benders decomposition method

$LB = -\infty, UB = \infty$
while $LB < UB - \epsilon$ **do**
    obtain a solution to the master problem
    get solution to master problem
    update the value of LB
    solve the sub-problem
    if the sub-problem is feasible **then**
        update UB
        generate an optimality cut
    else
        generate a feasibility cut
    end if
    Add cuts to the master problem
end while

6.4 Case Study

According to data from the National Institute of Public Health (INSP), in 2017, Mexico ranked seventh worldwide and third in Latin America in deaths from road accidents, with 22 deaths of young people between 15 and 29 years a day, and 24 thousand deaths per year on average. Road crashes are the leading cause of death in young people between 5 and 29 years of age and the fifth in the general population [20].

Mexico, like other developing countries, faces complex public health problems in the context of the growing demand for emergency services, mainly for injuries from external causes. Once an accident occurs, death, severe injury, and disability can be mitigated through timely and appropriate intervention by trained persons. In many cases, the speed of emergency care and the transfer of injured victims from the scene of an incident to a healthcare facility can save lives, reduce the incidence of disability in the short term, and significantly reduce the consequences.

Based on the government proposal [17] for the integration of the Model of Prehospital Medical Care, where it is considered as a specific objective to reduce prehospital care times, with the model proposed in this paper, it is intended to find an alternative solution for improving response times of prehospital services, so that the social impact generated by traffic accidents becomes lower.

Since most traffic accidents occur in urban areas [18] and the most congested city in the country is Mexico City [19], the latter was selected to test the functioning of the proposed model.

The main points where accidents occur are primary and secondary road cruises; these define the demand nodes or J set of the model. Sixteen accident zones were identified.

The candidate sites for the location of the bases for emergency vehicles are modules of citizen attention in charge of the government of Mexico City, which is in charge of the deployment of emergency services at the prehospital level. Seventy-one candidate sites (I) were identified.

Both sets of nodes are shown in the Figs. 6.1 and 6.2.

In this problem, two types of vehicles were considered: traditional ambulances and motorcycle ambulances. The care costs per type of vehicle were calculated based on [21]. The number of accidents per area per day, as well as their probability of occurrence, were calculated based on the data from [18]. Two possible occurrences per demand node were identified, and their probabilities were estimated. The resulting model was one with 65,536 scenarios.

Based on this information, the model was resolved, and the following results were obtained:

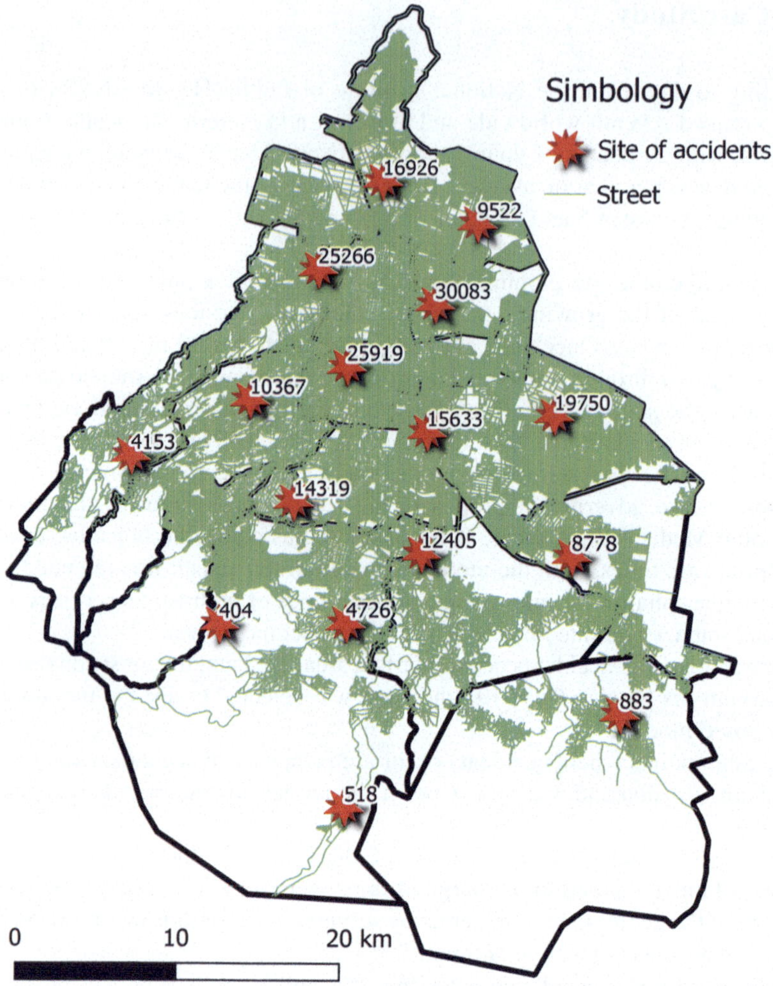

Fig. 6.1 Main intersections with accidents. Source: own preparation

- In order to cover 95% of the demand, five ambulance bases are required, located as shown in Fig. 6.3.
- The type of vehicles required per base are shown in Table 6.1.

6.5 Conclusions

The location of emergency vehicle bases and the assignment of services to points of demand is one of the most important aspects of health systems. In this work, we developed a model to locate bases for two types of ambulances and obtained

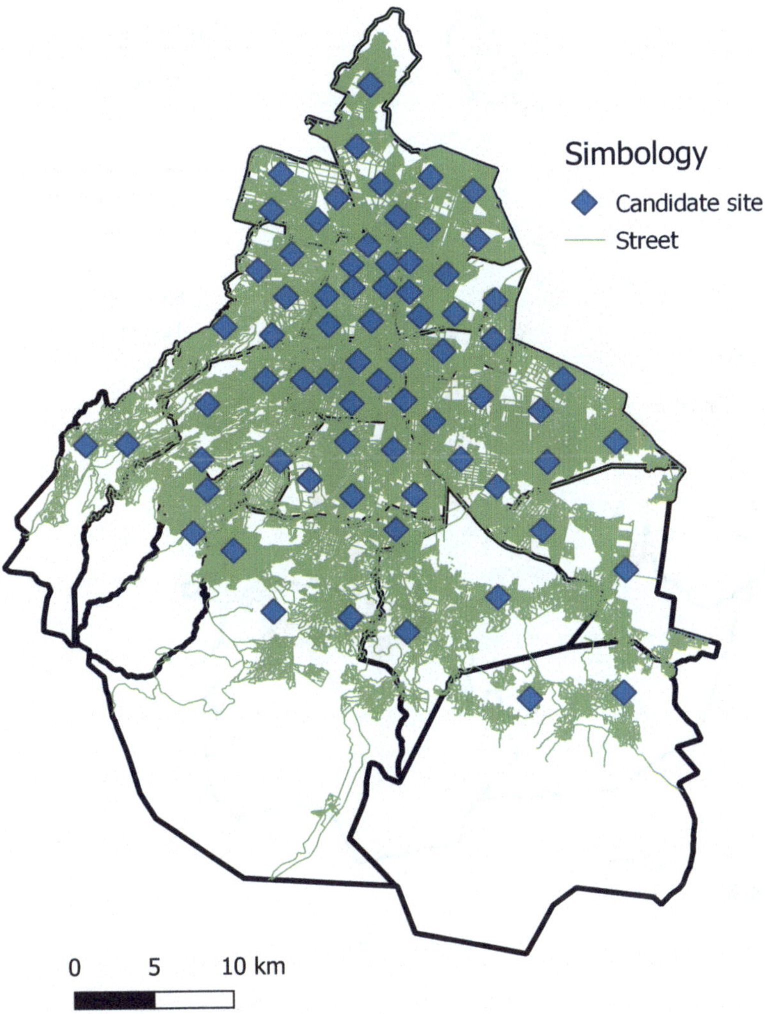

Fig. 6.2 Modules of citizen assistance. Source: own preparation

a proposal for allocation of services. We assumed that the occurrence of traffic accidents (demand) was uncertain, and used a decomposition method to resolve the resulting model.

We proposed a model for the problem location of a set of bases for emergency vehicles, as well as to decide the number of ambulances in them. The problem is solved to cover a region under stochastic demand, so it was considered a formulation indexed by scenarios. Also, different types of vehicles were considered. The goal was to minimize the total cost associated with installed bases and patient care. A coverage restriction was imposed based on the U.S. Emergency Medical Services Act of 95%.

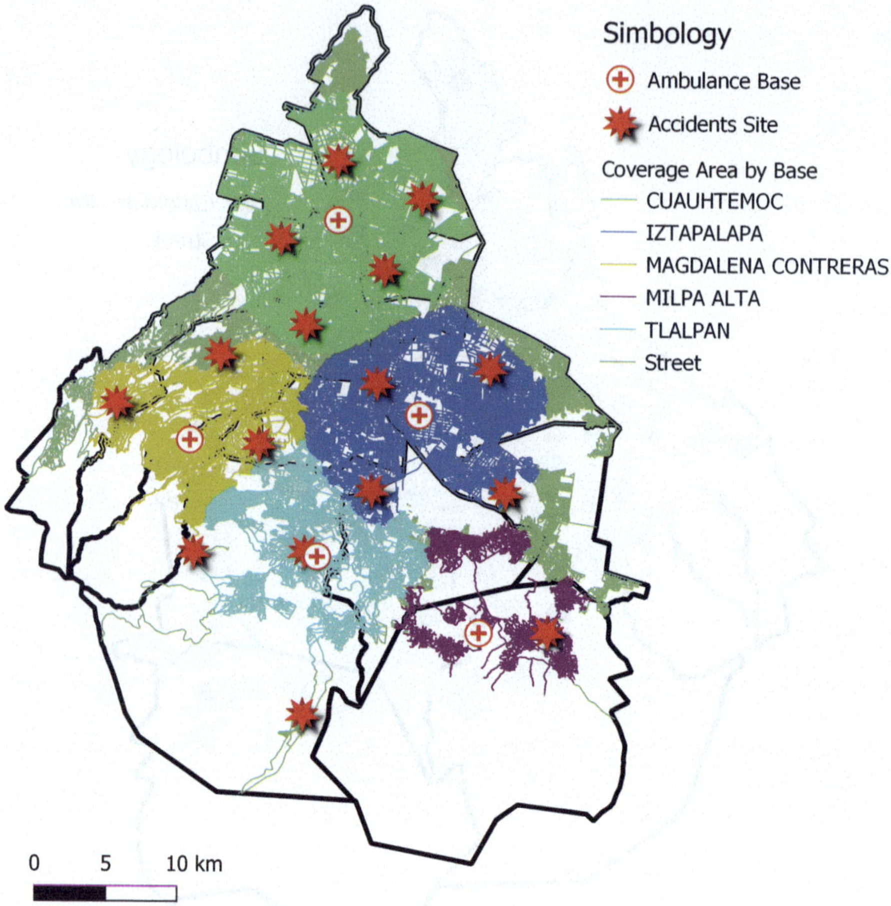

Fig. 6.3 Location of ambulance bases. Source: own preparation

Table 6.1 Vehicles required per base

Base	Traditional ambulances	Motorcycle ambulances
1	1	3
2	1	3
3	1	2
4	0	2
5	0	1

The operation of the model was developed and illustrated with an application in an urban area of Mexico City. The research was conducted considering the uncertainty of the demand, which makes the model very large and difficult to solve; however, the proposed solution method worked for the application shown. In practice, the number of scenarios in a problem of this type grows exponentially, which complicates the solution of the model using a global solver.

When calculating the expected value of perfect information and the value of the stochastic solution, we note that, even for such an instance, the relevance of using a stochastic approach is quite significant. Those aspects are a clear indication of the relevance of considering uncertainty in a problem such as the one presented.

References

1. M. Amiri-Aref, R. Farahani, M. Hewitt, W. Klibi, Equitable location of facilities in a region with probabilistic barriers to travel. Transp. Res. E Logist. Transp. Rev. **127**, 66–85 (2019). https://doi.org/10.1016/j.tre.2019.04.010
2. G. Benitez, G. Da Silveira, F. Fogliatto, Layout planning in healthcare facilities: a systematic review. Health Environ. Res. Des. J. **12**(3), 31–44 (2019). https://doi.org/10.1177/1937586719855336
3. D. Bertsimas, Y. Ng, Robust and stochastic formulations for ambulance deployment and dispatch. Eur. J. Oper. Res. **279**(2), 557–571 (2019). https://doi.org/10.1016/j.ejor.2019.05.011
4. H. Billhardt, M. Lujak, V. Sánchez-Brunete, A. Fernández, S. Ossowski, Dynamic coordination of ambulances for emergency medical assistance services. Knowl.-Based Syst. **70**, 268–280 (2014). https://doi.org/10.1016/j.knosys.2014.07.006
5. L. Brotcorne, G. Laporte, F. Semet, Ambulance location and relocation models. Eur. J. Oper. Res. **147**(3), 451–463 (2003). https://doi.org/10.1016/S0377-2217(02)00364-8
6. V. Bélanger, A. Ruiz, P. Soriano, Recent optimization models and trends in location, relocation, and dispatching of emergency medical vehicles. Eur. J. Oper. Res. (Mar 2018). https://doi.org/10.1016/j.ejor.2018.02.055
7. M.S. Daskin, L.K. Dean, *Location of Health Care Facilities* (Springer, Boston, 2004), pp. 43–76. https://doi.org/10.1007/1-4020-8066-2_3
8. S. Enayati, O. Özaltín, M. Mayorga, C. Saydam, Ambulance redeployment and dispatching under uncertainty with personnel workload limitations. IISE Trans. **50**(9), 777–788 (2018). https://doi.org/10.1080/24725854.2018.1446105
9. S. Firooze, M. Rafiee, S. Zenouzzadeh, An optimization model for emergency vehicle location and relocation with consideration of unavailability time. Sci. Iran. **25**(6E), 3685–3699 (2018). https://doi.org/10.24200/sci.2017.20022
10. A.M. Geoffrion, Generalized benders decomposition. J. Optim. Theory Appl. **10**(4), 237–260 (1972)
11. M. Labbé, M. Leal, J. Puerto, New models for the location of controversial facilities: a bilevel programming approach. Comput. Oper. Res. **107**, 95–106 (2019). https://doi.org/10.1016/j.cor.2019.03.003
12. C. Lei, W.H. Lin, L. Miao, A stochastic emergency vehicle redeployment model for an effective response to traffic incidents. IEEE Trans. Intell. Transp. Syst. **19**, 898–909 (2015). https://doi.org/10.1109/TITS.2014.2345480
13. X. Li, Z. Zhao, X. Zhu, T. Wyatt, Covering models and optimization techniques for emergency response facility location and planning: a review. Math. Methods Oper. Res. **74**(3), 281–310 (2011). https://doi.org/10.1007/s00186-011-0363-4. http://link.springer.com/10.1007/s00186-011-0363-4

14. M. Moeini, Z. Jemai, E. Sahin, An integer programming model for the dynamic location and relocation of emergency vehicles: a case study, in *Proceedings of the 12th International Symposium on Operational Research in Slovenia, SOR 2013* (2013), pp. 343–350
15. S. Nickel, M. Reuter-Oppermann, F. Saldanha-da Gama, Ambulance location under stochastic demand: a sampling approach. Oper. Res. Health Care **8**, 24–32 (2016). https://doi.org/10.1016/j.orhc.2015.06.006
16. M. Reuter-Oppermann, D. Richards, Decision support for EMS policy making using data analytics and real-time alerts, in *Proceedings - 2019 IEEE World Congress on Services, SERVICES 2019* (2019), pp. 266–271. https://doi.org/10.1109/SERVICES.2019.00079
17. C. Stconapra, Modelo de Atención Médica Prehospitalaria. Tech. Rep., Secretaría de Salud (2017)
18. Secretariado Técnico del Consejo Nacional para la Prevención de Accidentes (STCONAPRA). Informe Sobre la Situación de la Seguridad Vial, México, Technical report. Secretaría de Salud., 2017 (2016)
19. TomTom International BV. Traffic congestion ranking - TomTom Traffic Index 2017. Consulted in https://www.tomtom.com/en_gb/traffic-index/ranking/
20. S. Treviño, Consejos para prevenir los accidentes viales. Gaceta Instituto Nacional de Salud Pública **1**(1), 46–49 (2017)
21. D.A. Velez-Jaramillo, L.H. Lugo-Agudelo, B.C. Cano-Restrepo, P.A. Castro-García, H.I. García-García, Costos de atención y rehabilitación de pacientes con lesiones por accidentes de tránsito en el mundo. Revista Facultad Nacional de Salud Pública **34**(2), 220–229 (2016)
22. M. Zhang, Y. Zhang, Z. Qiu, H. Wu, Two-stage covering location model for air-ground medical rescue system. Sustainability **11**(12) (2019). https://doi.org/10.3390/su10023242
23. R. Zhang, B. Zeng, Ambulance Deployment with relocation through robust optimization. IEEE Trans. Autom. Sci. Eng. **16**(1), 138–147 (2019). https://doi.org/10.1109/TASE.2018.2859349

Chapter 7
Balance Layout Problem with the Optimized Distances Between Objects

S. Plankovskyy, A. Nikolaev, O. Shypul, I. Litvinchev, A. Pankratov, and T. Romanova

Contents

7.1 Introduction 85
7.2 Problem Formulation and Mathematical Model 86
7.3 Solution Strategy 89
7.4 Conclusions 91
References 92

7.1 Introduction

The layout problems have a wide spectrum of applications in science and technology, e.g., in assembling technological complexes [1], placing equipment on the aircraft board [2], and optimizing 3D printing [3, 4]. The problems take into account various constraints on the placement of objects in the domain (for example, the orientation of objects, the minimal allowed distances between objects, the presence of prohibited zones, balancing constraints [5–12]).

In this paper, we consider the problem of finding the layout of objects optimizing mutual distances between the objects and the distances between each object and the frontier of the domain. Constraints on the position of the gravity center and the moments of inertia of the system are also taken into account. Such problems

S. Plankovskyy · A. Nikolaev · O. Shypul
National Aerospace University named by M.Y. Zhukovskiy "Kharkiv Aviation Institute", Kharkiv, Ukraine

I. Litvinchev
Universidad Autónoma de Nuevo León, San Nicolás de los Garza, Mexico

A. Pankratov · T. Romanova (✉)
Department of Mathematical Modeling and Optimal Design, Institute for Mechanical Engineering Problems of the National Academy of Sciences of Ukraine, Kharkiv, Ukraine

P. Vasant et al. (eds.), *Data Analysis and Optimization for Engineering and Computing Problems*, EAI/Springer Innovations in Communication and Computing,
https://doi.org/10.1007/978-3-030-48149-0_7

arise, e.g., in adaptive manufacturing complex shape parts [13], in shock wave processes, such as ultrasonic hardening [14], thermopulse treatment by detonable gas mixtures [15].

7.2 Problem Formulation and Mathematical Model

Let a collection of three-dimensional arbitrary shaped canonically closed objects $T_i \subset R^3$, $i \in I = \{1, \ldots, N\}$ and a convex bounded set (domain) $Q = \left\{ (x, y, z) \in R^3 : \min_{k \in I_n} \{ f_k (x, y, z) \} \geq 0) \right\}$ be given, where each f_k is a differentiable function.

Objects T_i, $i \in I_N$ can be continuously translated and rotated. The motion vector of each object T_i is denoted by $u_i = (v_i, \theta_i)$, where $v_i = (x_i, y_i, z_i)$, $\theta_i = \left(\theta_{x_i}, \theta_{y_i}, \theta_{z_i}\right)$ for $i \in I_N$. The vector $u = (v, \theta)$ determines the layout of objects T_i, $i \in I_N$ in the Euclidean space R^3, where $v = (v_1, \ldots, v_N)$, $\theta = (\theta_1, \ldots, \theta_N)$.

The distance between objects $T_i(u_i)$ and $T_j(u_j)$ can be defined in the form

$$d_{T_i,T_j}\left(u_i, u_j\right) = \min_{t_i \in T_i(u_i), t_j \in T_j(u_j)} \| t_i - t_j \| .$$

The distance between object $T_i(u_i)$ and the frontier of Q is derived as follows:

$$d_{T_i,Q}(u_i) = \min_{t_i \in T_i(u_i), t \in Q^*} \| t_i - t \| ,$$

where $Q^* = R^3 \setminus \operatorname{int} Q$.

Let us introduce functions

$$F_1(u) = \frac{\sum\limits_{j=0}^{N-1} \sum\limits_{i=j+1}^{N} d_{ij}\left(u_i, u_j\right)}{0.5 \cdot N(N+1)},$$

$$F_2(u) = \frac{\sum\limits_{j=0}^{N-1} \sum\limits_{i=j+1}^{N} \left(d_{ij}\left(u_i, u_j\right) - F_1(u)\right)^2}{0.5 \cdot N(N+1)},$$

where $d_{ij}\left(u_i, u_j\right) = d_{T_i,T_j}\left(u_i, u_j\right)$, if $j > 0$, and $d_{ij}\left(u_i, u_j\right) = d_{T_i,Q}(u_i)$, $u_j = (0, 0, 0, 0, 0, 0)$, if $j = 0$.

Now we formulate the following balance layout problem:

find $u = (u_1, \ldots, u_N)$, such that

$$\begin{aligned} &F(u) = \alpha_1 F_1(u) - \alpha_2 F_2(u) \to \max, \\ &\alpha_1 + \alpha_2 = 1, \alpha_1 \geq 0, \alpha_2 \geq 0, \end{aligned} \tag{7.1}$$

providing non-overlapping $\mathrm{int}T_i(u_i) \cap \mathrm{int}\ T_j(u_j) = \varnothing,\ i > j \in I_N$, containment, $T_i(u_i) \subset Q,\ i \in I_N$, constraints and the balancing conditions discussed below.

First, we consider the optimization problem of the material point arrangement in the domain $Q \subset R^2$

$$\begin{aligned} &\kappa(u) \to \min, \\ &\kappa(u) = \tfrac{1}{\Gamma_Q} \int_{\Gamma_Q} \left((x_Q - x_1)^2 + (y_Q - y_1)^2 \right) d\Gamma_Q, \end{aligned} \tag{7.2}$$

where Γ_Q is the frontier of Q, L_Q is the length of Γ_Q.

Assuming the unit density of Γ_Q we have

$$\begin{aligned} &\int_{\Gamma_Q} x^2 d\Gamma_Q = J_y^Q, \quad \int_{\Gamma_Q} y^2 d\Gamma_Q = J_x^Q \\ &\int_{\Gamma_Q} x d\Gamma_Q = L_Q\ x_e^Q, \quad \int_{\Gamma_Q} y\ d\Gamma_Q = L_Q\ y_e^Q \end{aligned} \tag{7.3}$$

where J_x^Q, J_y^Q are axial moments of inertia of Γ_Q relative to the corresponding axis, and x_e^Q, y_e^Q are coordinates of the gravity center of Γ_Q.

We assume that the axes OX and OY pass through the gravity center of Γ_Q. Then, taking into account relations (7.3), the function (7.2) becomes

$$\kappa(u) = \frac{\left(J_x^Q + J_y^Q\right)}{L_Q} + \left(x_1^2 + y_1^2\right). \tag{7.4}$$

Considering that the sum of the axial moments of inertia of Γ_Q relative to the axes passing through its gravity center is a constant, it follows from (7.4) that $\min\ \kappa(u) = \frac{\left(J_x^Q + J_y^Q\right)}{L_Q}$ when $x_1 = y_1 \equiv 0$. Thus, the optimal layout of the material point according to (7.2) be at the gravity center of Γ_Q.

For layout of the canonically closed object $T_1 \subset R^2$ within the convex bounded domain $Q \subset R^2$ the function $\kappa(u)$ takes the form

$$\kappa(u) = \frac{1}{L_{T_1} L_Q} \int_{\Gamma_{T_1}} \left[\int_{\Gamma_Q} \left((x_Q - x_1)^2 + (y_Q - y_1)^2 \right) d\Gamma_Q \right] d\Gamma_{T_1}, \tag{7.5}$$

where Γ_{T_1} is the frontier of T_1 and L_{T_1} is the length of Γ_{T_1}.

It should be noted that

$$\min \kappa(u) = \frac{\left(J_x^Q + J_y^Q\right)}{L_Q} + \frac{\left(J_x^{T_1} + J_y^{T_1}\right)}{L_{T_1}}, \tag{7.6}$$

attains when the gravity centers of Γ_Q and Γ_{T_1} coincide.

The difference between the squares of the central radii of inertia of the domain Q and the object T_1 can also be used as the objective

$$r_Q^2 - r_1^2 = \frac{1}{2L_Q}\int\limits_{\Gamma_Q}\left(x_Q^2 + y_Q^2\right)\, d\Gamma_Q - \frac{1}{2L_{T_1}}\int\limits_{\Gamma_{T_1}}\left(x_1^2 + y_1^2\right)\, d\Gamma_{T_1}, \tag{7.7}$$

for which

$$\max\left(r_Q^2 - r_1^2\right) = \frac{\left(J_x^Q + J_y^Q\right)}{2L_Q} - \frac{\left(J_x^{T_1} + J_y^{T_1}\right)}{2L_{T_1}}. \tag{7.8}$$

However, criteria (7.5), (7.7) do not allow determining the orientation of the object T_1 relative to the OX and OY axes, since their extreme values (7.6), (7.8) do not depend on the rotation angle θ_1.

In this regard, instead of using $\left(r_Q^2 - r_1^2\right)$ as the optimization criteria, we will use the difference of the squares of the radii of inertia of Γ_Q and the object T_1 relative to the main central axes of inertia of Γ_Q.

We assume that the maximal main central moment of inertia $J_{\max}^Q$ corresponds to the moment of inertia relative to the OY axis, and the minimal main central moment of inertia $J_{\min}^Q$ corresponds to the moment of inertia relative to the OX axis.

Let us introduce the following requirements:

$$r_{\max\ Q}^2 - r_{y\ T_1}^2 = \frac{J_{\max}^Q}{L_Q} - \frac{\left(J_{\max}^{T_1}\cos^2\theta_1 + J_{\min}^{T_1}\sin^2\theta_1\right)}{L_{T_1}} \to \min, \tag{7.9}$$

$$r_{\min\ Q}^2 - r_{x\ T_1}^2 = \frac{J_{\min}^Q}{L_Q} - \frac{\left(J_{\min}^{T_1}\cos^2\theta_1 + J_{\max}^{T_1}\sin^2\theta_1\right)}{L_{T_1}} \to \max, \tag{7.10}$$

where $J_{\min}^{T_1}$, $J_{\max}^{T_1}$ are the main central moment of inertia of Γ_1.

If $\theta_1 = 0$ then

$$\min\left(r_{\max\ Q}^2 - r_{y\ T_1}^2\right) = \frac{J_{\max}^Q}{L_Q} - \frac{J_{\max}^{T_1}}{L_{T_1}}, \tag{7.11}$$

$$\max\left(r_{\min\ Q}^2 - r_{x\ T_1}^2\right) = \frac{J_{\min}^Q}{L_Q} - \frac{J_{\min}^{T_1}}{L_{T_1}}. \tag{7.12}$$

It leads to the requirement that the main central axes of the curves Γ_Q and Γ_{T_1} coincide.

Expressions (7.11) and (7.12) can be used to determine how to replace the object T_1 with a primitive object O_1. This requires the fulfillment of the following conditions

$$\min\left(r^2_{\max\ Q} - r^2_{y\ T_1}\right) \to \text{idem},\quad \max\left(r^2_{\min\ Q} - r^2_{x\ T_1}\right) \to \text{idem},$$

that implies

$$J^{O_1}_{\min} = \frac{J^{T_1}_{\min} \cdot L_{O_1}}{L_{T_1}};\quad J^{O_1}_{\max} = \frac{J^{T_1}_{\max} \cdot L_{O_1}}{L_{T_1}}.$$

In the three-dimensional case, these conditions should be formulated for the thin shells Λ_{T_1} and Λ_{O_1} stretched on the outer surfaces of the objects T_1 and O_1,

$$J^{O_1}_x = \frac{J^{T_1}_x \cdot S_{O_1}}{S_{T_1}};\quad J^{O_1}_y = \frac{J^{T_1}_y \cdot S_{O_1}}{S_{T_1}};\quad J^{O_1}_z = \frac{J^{T_1}_z \cdot S_{O_1}}{S_{T_1}}, \tag{7.13}$$

where OX, OY, OZ are the main central axes of Λ_{O_1} and Λ_{T_1}, S_{O_1} and S_{T_1} are their areas.

Taking into account (7.13), mathematical model of the layout problem, using the phi-function technique [6, 7, 16], can be represented in the following form:

$$F(u) = \alpha_1 F_1(u) - \alpha_2 F_2(u) \to \max_{(u,\tau)\in R^\sigma}, \tag{7.14}$$

$$W = \left\{(u,\tau) : \Phi'_{ij}\left(u_i, u_j, \tau_{ij}\right) \geq 0,\ \ i < j \in I_N, \Phi_i\left(u_i\right) \geq 0, i \in I_N, \Upsilon(u) \geq 0\right\}, \tag{7.15}$$

$\Phi'_{ij}\left(u_i, u_j, \tau_{ij}\right)$ is a normalized quasi phi-function of objects $O_i(u_i)$ and $O_j(u_j)$, $i < j \in I_N$ [16]; $\Phi_i(u_i)$ is a normalized phi-function of objects $O_i(u_i)$ and Q^* [6]; $\tau = (\tau_{11}, \tau_{12}, \ldots, \tau_{ij}, \ldots, \tau_{(N-1),N})$, τ_{ij} is a vector of auxiliary variables for quasi phi-function $\Phi'_{ij}\left(u_i, u_j, \tau_{ij}\right)$; $\Phi'_{ij}\left(u_i, u_j, \tau_{ij}\right) \geq 0$ is responsible for the object non-overlapping; $\Phi_i(u_i) \geq 0$ is responsible for containment of the object into the domain; $\Upsilon(u) \geq 0$ is responsible for the balancing conditions of the system.

The function $d_{ij}(u_i, u_j)$ in $F_1(u)$,$F_2(u)$ is defined for problem (7.14)–(7.15) as following: $d_{ij}\left(u_i, u_j\right) = \max\left\{0, \max\limits_{\tau_{ij}} \Phi'_{ij}\left(u_i, u_j, \tau_{ij}\right)\right\}$, if $j > 0$, and $d_{ij}(u_i, u_j) = \max\{0, \Phi_i(u_i)\}$, $u_j = (0, 0, 0, 0, 0, 0)$, if $j = 0$.

7.3 Solution Strategy

To solve this problem, the following strategy is proposed.

At the first stage, using the formulas (7.13), the shape of the primitive objects $(O_i,\ i \in I_N)$ is determined. As an additional requirement for choosing the shape of the objects $(O_i,\ i \in I_N)$, the condition for maintaining the volume of placed objects ($O_i, i \in I_N$) with the volume of original objects (T_i, $i \in I_N$) is used (i.e., $V(O_i) = V(T_i)$).

At the second stage, the auxiliary problem of placing primitive objects O_i in an optimized domain $Q_\mu = \mu Q$ is solved

$$\mu \to \min, \quad (u, \tau, \mu) \in V \subset R^{\sigma+1}, \tag{7.16}$$

$$V = \Big\{ (u, \tau, \mu) : \Phi'_{ij}\left(u_i, u_j, \tau_{ij}\right) \geq 0, \;\; i < j \in I_N, \\ \Phi_i\left(u_i, \mu\right) \geq 0, i \in I_N, \Upsilon\left(u, \mu\right) = 0 \Big\} \tag{7.17}$$

where $\mu > 0$ is a homothetic coefficient of Q, $\Phi_i(u_i, \mu)$ is a phi-function of objects $O_i(u_i)$ and Q^*_μ.

In expression (7.17), the equation $\Upsilon(u, \mu) = 0$ describes the condition of coincidence of the gravity centers of the thin shell Λ_{Q_μ} and the system of thin shells, $i \in I_N$. When placing the origin of the global coordinate system in the gravity center of the shellΛ_{Q_μ} , the function $\Upsilon(u)$ has the form

$$\Upsilon(u) = \frac{\sum\limits_{i=1}^{N} S_{\Lambda_{O_i}} \overrightarrow{R}_{\Lambda_{O_i}}}{\sum\limits_{i=1}^{N} S_{\Lambda_{O_i}}}, \tag{7.18}$$

where $\overrightarrow{R}_{\Lambda_{O_i}}$ is the radius vector of the gravity center of the thin shell Λ_{O_i}. A local extremum point of problem (7.16)–(7.18) is denoted by (u^*, τ^*, μ^*).

At the third stage, the following subproblems are solved:

- the resulting arrangement of objects $O_i\left(u_i^*\right)$, $i \in I_N$ is oriented in the domain Ω so that the gravity centers of the thin shell Λ_Ω and the system of thin shells Λ_{O_i}, $i \in I_N$, coincide, as well as, the directions of their main central axes of inertia. Getting the point u^{**};
- translate objects $O_i(u_i)$, $i \in I_N$ in the direction of vectors $\overrightarrow{R}_{\Lambda_{O_i}}$, $i \in I_N$ is carried out with a coefficient λ. The coefficient is determined by solving the following auxiliary problem:

$$F\left(\lambda v, \theta, \tau\right) \to \max_{\lambda},$$

starting from the point (u^{**}, τ^{**}), $u^{**} = (v^{**}, \theta^{**})$ with fixed parameters θ^{**}, τ^{**}. Get the point (u^{***}, τ^{**}), $u^{***} = (v^{***}, \theta^{**})$.

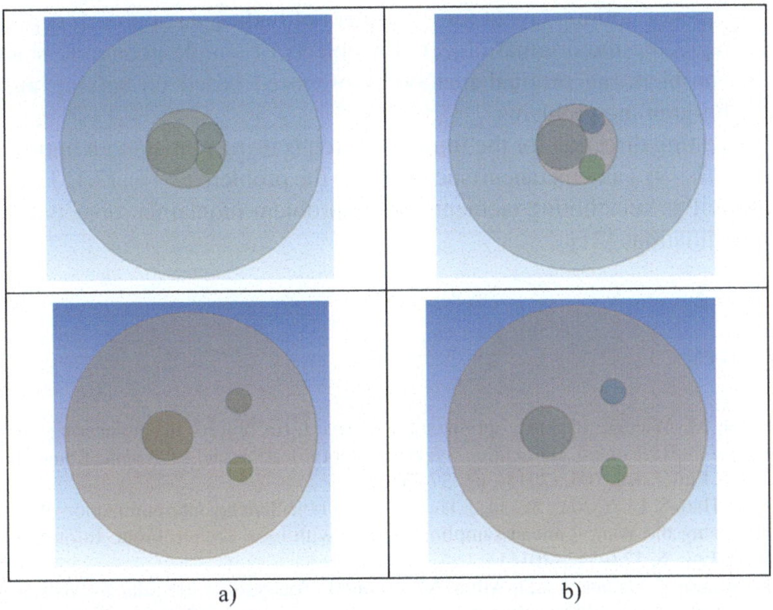

Fig. 7.1 Illustration to the solution strategy: column (**a**) using the first algorithm; column (**b**) using the second algorithm

– the objects $O_i(u^{***})$, $i \in I_N$ are replaced with the original objects $T_i(u^{***})$, $i \in I_N$ while maintaining the direction of the main central axes of inertia of the shells Λ_{O_i} and Λ_{T_i}.

To solve a test layout problem for primitive objects $O_i(u_i)$, $i = 1,2,3$ two heuristic algorithms are used: (1) solving the packing problem (7.16)–(7.17) with the subsequent translation of the objects; (2) solving the packing problem taking into account the additional restriction (7.18) with the subsequent translation of the objects.

Figure 7.1 illustrates the results of the arrangement of primitive objects $O_i(u_i)$, $i = 1,2,3$ using two proposed algorithms.

7.4 Conclusions

The problem of optimal layout subject to a uniform remoteness of objects from the border of the region is considered. It is shown that the optimal placement of a single object is achieved when the centers of gravity of the thin shells corresponding to the surfaces of the object and the area of placement, as well as the directions of their main central axes of inertia, coincide.

For the case of optimal layout for the system of bodies, an approach is proposed based on replacing the original objects for objects of simple geometric shape. To solve this problem, an original method is proposed based on solving auxiliary nonlinear programing problems.

An interesting direction for the future research is using non-smooth optimization (see, e.g., [17–19].) and metaheuristics to solve the problem (7.14), (7.15) (see, e.g., [20]), as well as substituting elements of the problem (container, objects) by their grid approximations [21].

References

1. G. Fadel, M. Wiecek, Packing optimization of free-form objects in engineering design, in *Optimized Packings with Applications*, Springer Optimization and Its Applications, ed. by G. Fasano, J. Pintér, vol. 105, (2015), pp. 37–66
2. J. Liu, L. Hao, S. Li, Y. Xue, Z. Liu, J. Huang, Multi-objective layout optimization of a satellite module using the Wang-Landau sampling method with local search. Front. Inform. Technol. Electron. Eng. **6**, 527–542 (2016)
3. L.J.P. Araújo, E. Özcan, J.A.D. Atkin, M. Baumers, Analysis of irregular three-dimensional packing problems in additive manufacturing: a new taxonomy and dataset. Int. J. Prod. Res. **57**(18), 5920–5934 (2018)
4. J. Gardan, Additive manufacturing technologies: state of the art and trends. Int. J. Prod. Res. **54**(10), 3118–3132 (2016)
5. A. Bortfeldt, G. Wäscher, Constraints in container loading: a state-of-the-art review. Eur. J. Oper. Res. **229**(1), 1–20 (2013)
6. Y. Stoyan, T. Romanova, Mathematical models of placement optimization: two- and three-dimensional problems and applications, in *Modeling and Optimization in Space Engineering*, Springer Optimization and Its Applications, ed. by G. Fasano, J. Pintér, vol. 73, (2013), pp. 363–388
7. Y. Stoyan, A. Pankratov, T. Romanova, A. Chugay, Optimized object packings using quasi-phi-functions, in *Optimized Packings and Their Applications*, Springer Optimization and Its Applications, ed. by G. Fasano, J. Pintér, vol. 105, (Springer, New York, 2015), pp. 265–291
8. G. Fasano, Global optimisation point of view for non-standard packing problems. J. Glob. Optim. **55**(2), 279–299 (2013)
9. S.X. Li, J. Zhao, P. Lu, Y. Xie, Maximum packing densities of basic 3D objects. Chinese Sci. Bull. **55**(2), 114–119 (2010)
10. X. Liu, J. Liu, A. Cao, Z. Yao, HAPE3D – a new constructive algorithm for the 3D irregular packing problem. Front. Inform. Technol. Electron. Eng. **16**(5), 380–390 (2015)
11. M. Verkhoturov, A. Petunin, G. Verkhoturova, K. Danilov, D. Kurennov, The 3D object packing problem into a parallelepiped container based on discrete-logical representation. IFAC-PapersOnLine **49**(12), 1–5 (2016)
12. T. Romanova, J. Bennell, Y. Stoyan, A. Pankratov, Packing of concave polyhedra with continuous rotations using nonlinear optimization. Eur. J. Oper. Res. **268**, 37–53 (2018)
13. D.H. Zhang, Y. Zhang, B.H. Wu, Research on the adaptive machining technology of blisk. Adv. Mater. Res. **69-70**, 446–450 (2009)
14. M. Rakita, M. Wang, Q. Han, Y. Liu, F. Yin, Ultrasonic shot peening. Int. J. Comput. Mater. Sci. Surf. Eng. **5**(3), 189–209 (2013)
15. S. Plankovskyy, A. Teodorczyk, O. Shypul, O. Tryfonov, D. Brega, Determination of detonable gas mixture heat fluxes at thermal deburring. Acta Politech. **59**(2), 162–169 (2019)

16. Y. Stoyan, A. Pankratov, T. Romanova, Quasi-phi-functions and optimal packing of ellipses. J. Glob. Optim. **65**(2), 283–307 (2016)
17. A.A. Kovalenko, T.E. Romanova, P.I. Stetsyuk, Balance layout problem for 3D-objects: mathematical model and solution methods. Cybern. Syst. Anal. **51**, 556–565 (2015)
18. P. Stetsyuk, T. Romanova, G. Scheithauer, On the global minimum in a balanced circular packing problem. Optim. Lett. **10**, 347–1360 (2016)
19. Y. Stoyan, A. Pankratov, T. Romanova, G. Fasano, J. Pintér, Y. Stoian, A. Chugay, Optimized packings in space engineering applications: part I, in *Modeling and Optimization in Space Engineering State of the Art and New Challenges*, ed. by G. Fasano, J. D. Pinter, (Springer, New York, 2019), pp. 395–437
20. R. Torres-Escobar, J.A. Marmolejo-Saucedo, I. Litvinchev, Binary monkey algorithm for approximate packing non-congruent circles in a rectangular container. Wirel. Netw. (2018). https://doi.org/10.1007/s11276-018-1869-y
21. I. Litvinchev, L. Infante, L. Ozuna, Approximate circle packing in a rectangular container: integer programming formulations and valid inequalities, in *Lecture Notes in Computer Science, 8760/2014*, (2014), pp. 47–60

Chapter 8
Optimized Packing of Object Clusters with Balancing Conditions

T. Romanova, A. Pankratov, I. Litvinchev, and Jose Antonio Marmolejo-Saucedo

Contents

8.1 Introduction 95
8.2 Problem Statement 96
8.3 Mathematical Modeling 98
8.4 Solution Algorithm 99
8.5 Computational Results 101
8.6 Conclusions 106
References 106

8.1 Introduction

A two dimensional packing problem is proven to be NP-hard [1]. Along with traditional logistics applications, this problem frequently arises in industrial engineering, project portfolio selection, production planning, material science, and physics (see, e.g., [2] and the references therein).

Different objects and containers are considered in the related papers. In particular, packing ellipses is presented in [3–8], circular packing is studied in [9–14], packing convex polygons is presented in [15–18], irregular packing for arbitrary shaped objects is studied in [19–24].

Frequently the objects have to form a number of clusters of non-overlapping objects. For example, in container loading problem [25] the objects are clustered according to their shapes, types, brands, or colors. The objects corresponding to the

T. Romanova · A. Pankratov
Department of Mathematical Modeling and Optimal Design, Institute for Mechanical Engineering Problems of the National Academy of Sciences of Ukraine, Kharkiv, Ukraine

I. Litvinchev
Universidad Autónoma de Nuevo León, San Nicolás de los Garza, Mexico

J. A. Marmolejo-Saucedo (✉)
Universidad Panamericana, Facultad de Ingeniería, Ciudad de México, México
e-mail: jmarmolejo@up.edu.mx

P. Vasant et al. (eds.), *Data Analysis and Optimization for Engineering and Computing Problems*, EAI/Springer Innovations in Communication and Computing,
https://doi.org/10.1007/978-3-030-48149-0_8

same supplier or a final destination (client) also can be grouped into clusters to help the decision maker loading/unloading the container.

The number of objects to be placed in the cluster is frequently fixed together with their shapes. However, the overall shape of the cluster in many cases is not specified and depends on the objects layout inside the cluster. The shape of the cluster is defined in [26] as a convex hull of the objects that form the cluster. Assuming the objects are non-overlapping, the shape of the cluster is changed depending on the placement of moving objects in the cluster.

In this paper clusters composed of the same shapes are considered. However, different sizes are permitted for objects in the same cluster. It is assumed that the number of clusters, number of objects in the cluster as well as their shapes, sizes, and masses are given. A packing for clusters is said to be feasible if: (a) objects in the cluster are non-overlapping; (b) clusters are non-overlapping and belong completely to a rectangular container, and (c) balancing conditions are taken into account. To get the sparse clusters layout a specific "distance" between the clusters is maximized. This approach arises from the requirements to have more space between clusters to simplify their loading/unloading. In contrast to packing problem considered in [26], in this paper balancing conditions are taking into account. Balancing (equilibrium) conditions are important in container loading problems. They are imposed to prevent container rotation during portal movements.

Now we briefly note the novelty of the presented manuscript: (1) new problem formulation for the optimized packing clusters subject to balancing conditions; (2) novel nonlinear programming model for the optimized equilibrium packing clusters; (3) new fast approach for getting starting feasible points to search for local optimal solutions.

8.2 Problem Statement

Let Ω be a rectangular container of the given length l, width w, and mass w_0. And let $T_i, i \in \{1, 2, \ldots, n\} = I_n$, be an ordered collection of convex objects. Each object T_i is defined by its metrical characteristics (sizes) and its mass w_i. To characterize position of the object T_i a vector of variable placement parameters$u_i = (v_i, \theta_i)$ is used. Here $v_i = (x_i, y_i)$ is a translation vector and θ_i is a rotation angle. It is assumed that the object is centered at the origin of its own coordinate system. Displacement of the object T_i, i.e. its rotation by an angle θ_i and translation by a vector v_i can be expressed as $T_i\,(u_i) = \left\{p \in R^2 : p = v_i + M\,(\theta_i) \cdot p^0, \forall p^0 \in T_i^0\right\}$, where T_i^0 denotes non-translated and non-rotated object T_i, while $M(\theta_i)$ is a rotation matrix.

We denote the center of mass of the container by (x_0, y_0). Let the center of mass of the object T_i be placed at $v_i = (x_i, y_i)$ for $i \in I_n$. We designate a system of objects T_i, $i = 1, \ldots, n$, placed into container Ω by Ω_A.

The expected position of the mass center of Ω_A is defined as

$$x_s(v) = \frac{1}{M}\sum_{i=0}^{n} w_i x_i, \quad y_s(v) = \frac{1}{M}\sum_{i=0}^{n} w_i y_i, \quad M = \sum_{i=0}^{n} w_i,$$

where $v = (v_1, v_2, \ldots, v_n)$.

Let $(\Delta x_e, \Delta y_e)$ be allowable deviations from the given point $(x_e, y_e) \in \Omega$.

A cluster of objects (see [26]) involves a collection of objects that have the same configuration. The set of objects T_i, $i = 1, \ldots, n$, is splitted into clusters $\Lambda_1, \Lambda_2, \ldots, \Lambda_N$ according to the given index sets $\Xi_1 = \{1, \ldots, n_1\}$, $\Xi_2 = \{n_1 + 1, \ldots, n_2\}$, $\ldots$, $\Xi_N = \{n_{N-1} + 1, \ldots, n\}$, $n = \sum_{q=1}^{N} m_q$, where $m_q = card\ (\Xi_q)$ for $q \in \{1, \ldots, N\} = J_N$, i.e. $\Lambda_1 = \{T_1, \ldots, T_{n_1}\}$, $\Lambda_2 = \{T_{n_1+1}, \ldots, T_{n_2}\}$, $\ldots$, $\Lambda_N = \{T_{n_{N-1}+1}, \ldots, T_n\}$.

The vectors of design parameters of clusters $\Lambda_1, \Lambda_2, \ldots, \Lambda_N$ are denoted by $z_1 = (u_1, \ldots, u_{n_1})$, $z_2 = (u_{n_1+1}, \ldots, u_{n_2})$, $\ldots$, $z_N = (u_{n_{N-1}+1}, \ldots, u_n)$.

Further a notation $\Lambda_q(z_q)$ is used to denote a cluster Λ_q containing objects $T_i(u_i)$, $i \in \Xi_q$, i.e. $\Lambda_q(z_q) = \bigcup_{i \in \Xi_q} T_i(u_i)$.

The convex hull of objects $T_i(u_i)$, $i \in \Xi_q$, denoted by $\widehat{\Lambda}_q(z_q) = conv \Lambda_q(z_q)$, is associated with each cluster $\Lambda_q(z_q)$ for $q \in J_N$.

According to [26], (1) two clusters $\Lambda_q(z_q)$ and $\Lambda_g(z_g)$ are said to be non-overlapping if their convex hulls do not overlap, (2) a cluster $\Lambda_q(z_q)$ is in the container Ω if so its convex hull.

Cluster Packing Problem with Balancing Conditions (CLPB). Arrange the collection of clusters $\Lambda_q(z_q)$, $q \in J_N$, sufficiently distant one from another into a rectangular container Ω taking into account balancing conditions, such that

$$\text{int}\, T_i(u_i) \cap \text{int}\, T_j(u_j) = \varnothing, \quad \text{for } i > j, (i, j) \in \Sigma_q \times \Sigma_q, \ q \in J_N, \tag{8.1}$$

$$\text{int}\, \widehat{\Lambda}_q(z_q) \cap \text{int}\, \widehat{\Lambda}_g(z_g) = \varnothing, \quad \text{for } q > g \in J_N, \tag{8.2}$$

$$\widehat{\Lambda}_q(z_q) \subset \Omega \quad \text{for each } q \in I_N. \tag{8.3}$$

The balancing conditions are defined as follows:

$$\mu(v) = \min\{\mu_1(v), \mu_2(v)\} \geq 0, \tag{8.4}$$

where

$$\mu_1(v) = \min\{-(x_s(v) - x_e) + \Delta x_e, (x_s(v) - x_e) + \Delta x_e\},$$
$$\mu_2(v) = \min\{-(y_s(v) - y_e) + \Delta y_e, (y_s(v) - y_e) + \Delta y_e\}.$$

The constraint (8.1) stands for non-overlapping objects in the same cluster, the constraint (8.2) presents non-overlapping clusters, the constraint (8.3) assures containment clusters into the container Ω, while the relation (8.4) describes balancing conditions.

8.3 Mathematical Modeling

Sparse packing clusters in the container Ω can be stated as follows:

$$\max_{(u,\phi,\psi,\rho)\in W\subset R^{\sigma}} \rho, \tag{8.5}$$

$$W = \Big\{ (u,\phi,\psi,\rho) \in R^{\sigma} : \Phi'^{\widehat{\Lambda}_q \widehat{\Lambda}_g g}\left(z_q, z_g, \varphi_{qg}, \gamma_{qg}\right) - 0.5\rho \geq 0, q > g \in I_N, \rho \geq 0,$$
$$\Phi'^{T_i T_j}\left(u_i, u_j, u_{ij}\right) \geq 0, i > j, (i,j) \in \Xi_q \times \Xi_q, q \in I_N,$$
$$\Phi^{T_i \Omega^*}\left(u_i\right) \geq 0, i \in I_n, \mu(v) \geq 0 \Big\}. \tag{8.6}$$

Here $u = (u_1, u_2, \ldots, u_n)$, $u_i = (v_i, \theta_i)$, $v_i = (x_i, y_i)$, ρ stands for the distance between $\widehat{\Lambda}_q\left(z_q\right)$ and $\widehat{\Lambda}_g\left(z_g\right)$, $\psi = (\varphi_{qg}, \gamma_{qg}, q > g \in J_N)$ stands for auxiliary variables in the quasi phi-function $\Phi'^{\widehat{\Lambda}_q \widehat{\Lambda}_g g}\left(z_q, z_g, \varphi_{qg}, \gamma_{qg}\right)$ for $\widehat{\Lambda}_q\left(z_q\right)$ and $\widehat{\Lambda}_g\left(z_g\right)$, $\phi = \left(u_{ij}, (i,j) \in \Xi_q \times \Xi_q, q \in I_N\right)$ is the vector of auxiliary variables in the quasi phi-function $\Phi'^{T_i T_j}\left(u_i, u_j, u_{ij}\right)$for $T_i(u_i)$ and $T_j(u_j)$, $\Phi^{T_i \Omega^*}\left(u_i\right)$ is the phi-function for $T_i(u_i)$ and Ω^*. Explicit expressions for the (quasi) phi-functions one can find in [26]. The details on Stoyan's phi-function technique are provided in, e.g., [27–29]. Here $\sigma = 1 + 3n + \frac{\tau_q}{2} \sum_{q=1}^{N} m_q\left(m_q - 1\right) + N\left(N - 1\right)$ is the number of decision variables in (8.5)–(8.6), τ_q is a number of auxiliary variables, $(i,j) \in \Xi_q \times \Xi_q$, $q \in I_N$ and $\mu(v) \geq 0$ is the balancing condition defined in (8.4).

The nonlinear programming problem (8.5)–(8.6) gives all optimal solutions to the packing problem (an exact formulation).

The problem (8.5)–(8.6) has $O\left(\sum_{q=1}^{N} m_q^2\right) + O\left(N^2\right)$ nonlinear inequalities and $O\left(\sum_{q=1}^{N} m_q^2\right) + O\left(N^2\right)$variables taking into account the auxiliary variables in quasi phi-functions.

The next section presents a solution technique for finding a local solution to the problem (8.5)–(8.6).

8.4 Solution Algorithm

The solution strategy highlighted in [26] is used. A multistart algorithm is used having the following stages: Stage 1. Construct a set of feasible starting points for the problem (8.5)–(8.6); Stage 2. Find a set of local maxima for (8.5)–(8.6) using a feasible point obtained at Stage 1; Stage 3. Select the best local maxima found at Stage 2.

An algorithm *ECSP* (*E*quilibrium *C*luster *S*tarting *P*oints) is proposed to generate starting points feasible to the problem (8.5)–(8.6). This algorithm involves the following Steps.

Step 1. For each cluster $\Lambda_q(z_q)$ a circular region λC_q is associated having variable center point $v_q = (x_q, y_q)$ and variable radius λR_q, where λ is a scaling parameter, $R_q = \sqrt{\sum_{i \in \Xi_q} S_i}$. Here S_i is the area of the object T_i, $i \in \Xi_q$, $q \in I_N$.

Step 2. Solve the problem of finding variable center points $v_q = (x_q, y_q)$ of the appropriate circular regions λC_q for $\lambda = 0$, $q \in I_N$, taking into account the balancing condition (8.4) with the cluster mass$w_q^c = \sum_{i=1}^{n_q} w_i$ for $q \in I_N$,and $x_s(v) = \frac{1}{M} \sum_{q=0}^{N} w_q^c x_q$, $y_s(v) = \frac{1}{M} \sum_{q=0}^{N} w_q^c y_q$, $w_0^c = w_0$.

$$\min_{(v,\xi) \in V} \xi, \tag{8.7}$$

$$V = \left\{ (v, \xi) \in R^{2N+1} : \varphi_q \left(v_q \right) \geq 0, q \in I_N, \xi - \mu^c(v) \geq 0 \right\}, \tag{8.8}$$

$$\begin{aligned}
&\mu^c(v) = \min \left\{ \mu_1^c(v), \mu_1^c(v) \right\}, \\
&\mu_1^c(v) = \min \left\{ - \left(x_s^c(v) - x_e \right) + \Delta x_e, \left(x_s^c(v) - x_e \right) + \Delta x_e \right\}, \\
&\mu_2^c(v) = \min \left\{ - \left(y_s^c(v) - y_e \right) + \Delta y_e, \left(y_s^c(v) - y_e \right) + \Delta y_e \right\}, \\
&\varphi_q \left(v_q \right) = \min_{s=1,\ldots,4} f_{is} \left(v_i \right) \\
&f_{i1} \left(v_i \right) = x_i, \; f_{i2} \left(v_i \right) = y_i, \; f_{i3} \left(v_i \right) = -x_i + l, \; f_{i4} \left(v_i \right) = -y_i + w,
\end{aligned} \tag{8.9}$$

Starting values of $v_q = (x_q, y_q) \in \Omega$ to solve problem (8.7)–(8.9) are generated randomly. Denote by $\left(v^0 = \left(v_1^0, \ldots, v_N^0 \right), \xi^0 \right)$ a local solution of the problem (8.7)–(8.9).

Step 3. Starting from the point $(v^0, \lambda^0 = 0)$, solve the sub-problem:

$$\max_{(v,\lambda)\in V} \lambda, \tag{8.10}$$

$$V = \Big\{ (v, \lambda) \in R^{2N+1} : \Phi^{C_q C_g}\left(v_q, v_g, \lambda\right) \geq 0, q > g \in I_N, \\ \Phi^{C_q \Omega^*}\left(v_q, \lambda\right) \geq 0, q \in I_N, \lambda \geq 0, \mu^c(v) \geq 0 \Big\}, \tag{8.11}$$

where $\Phi^{C_q C_g}\left(v_q, v_g, \lambda\right)$ is the phi-function for λC_q and λC_g with appropriate radii λR_q and λR_g and centers $v_q = (x_q, y_q)$ and $v_g = (x_g, y_g)$ for $q > g \in I_N$. Here $\Phi^{C_q \Omega^*}\left(v_q, \lambda\right)$ is a phi-function for a circle λC_q centered at $v_q = (x_q, y_q)$ and the object $\Omega^* = R^2 \backslash \mathrm{int}\Omega$, $q \in I_N$. The function $\mu^c(v)$ is defined in (8.9). A local solution to (8.10)–(8.11) is denoted by (v^*, λ^*).

Step 4. For each circular area $\lambda^* C_q$ generate randomly translations vectors $v_i = (x_i, y_i) \in \lambda^* C_q$ of objects $T_i(v_i)$, $i \in \Xi_q$, from the cluster $\Lambda_q(z_q)$, $q \in I_N$.

Step 5. For each Λ_q, $q \in I_N$, form a vector $z_q = \left(v_{n_{q-1}+1}, \ldots, v_{n_q}\right)$ and find the minimum deviation of the center mass $\left(x_s^q\left(z^q\right), y_s^q\left(z^q\right)\right)$ of the cluster from the point v_q^0 with the cluster mass $M_q = \sum_{i \in \Xi_q} w_i$ and $x_s^q\left(z^q\right) = \frac{1}{M_q} \sum_{i \in \Xi_q} w_i x_i$, $y_s^q\left(z^q\right) = \frac{1}{M_q} \sum_{i \in \Xi_q} w_i y_i$, solving the following NLP sub-problem:

$$\min_{(z^q, \xi^q)\in V^q} \xi^q, \tag{8.12}$$

$$V^q = \left\{ \left(z^q, \xi^q\right) \in R^{2n_q+1} : \varphi_i\left(v_i\right) \geq 0, i \in \Xi_q, \xi^q - \mu^q\left(z^q\right) \geq 0 \right\}, \tag{8.13}$$

$$\begin{aligned} &\mu^q\left(z^q\right) = \min\left\{\mu_1^q\left(z^q\right), \mu_1^q\left(z^q\right)\right\}, \\ &\mu_1^q\left(z^q\right) = \min\left\{-\left(x_s^q\left(z^q\right) - x_q^0\right) + \Delta x_e, \left(x_s^C\left(z^q\right) - x_q^0\right) + \Delta x_e\right\}, \\ &\mu_2^q\left(z^q\right) = \min\left\{-\left(y_s^q\left(z^q\right) - y_q^0\right) + \Delta y_e, \left(y_s^C\left(z^q\right) - y_q^0\right) + \Delta y_e\right\}, \\ &\varphi_i^q\left(v_i^q\right) = \left(\lambda^* R^q\right)^2 - \left(x_q^* - x_i\right)^2 - \left(y_q^* - y_i\right)^2. \end{aligned} \tag{8.14}$$

Let a local maximum for the problem (8.12)–(8.14) be (z^{q*}, ξ^{q*}).

Step 6. Initial values of $\varphi_{qg}^0, \gamma_{qg}^0$ are obtained by straightforward geometrical calculations for the quasi phi-function $\Phi'^{\Lambda_q \Lambda_g}\left(z_q, z_g, \varphi_{qg}, \gamma_{qg}\right), q > g \in I_N$ in the problem (8.5)–(8.6). These values are used to define separating lines between circular areas $\lambda^* C_q$ and $\lambda^* C_g$ of radii $\lambda^* R_q$ and $\lambda^* R_g$ with center points v_q^* and $v_g^*, q > g \in I_N$.

Step 7. Feasible points for the problem (8.5)–(8.6) are obtained from the following NLP problem:

$$\max_{(u,\phi,\psi,\beta)\in G} \beta, \tag{8.15}$$

$$G=\Big\{ (u,\phi,\psi,\beta) \in R^{\sigma} : \Phi'^{\widehat{\Lambda_q}\widehat{\Lambda_{gg}}}\left(z_q, z_g, \varphi_{qg}, \gamma_{qg}, \beta\right) \geq 0, q>g \in I_N, 0 \leq \beta \leq 1,$$
$$\Phi'^{T_iT_j}\left(u_i, u_j, u_{ij}, \beta\right) \geq 0, i > j, (i,j) \in \Xi_q \times \Xi_q, q \in I_N,$$
$$\Phi^{T_i\Omega^*}\left(u_i, \beta\right) \geq 0, i \in I_n, \mu(v) \geq 0\}\}, \tag{8.16}$$

where $u = (u_1, u_2, \ldots, u_n)$, $u_i = (v_i, \theta_i)$, $v_i = (x_i, y_i)$, $\psi = (\varphi_{qg}, \gamma_{qg}, q > g \in J_N)$; β is a scaling coefficient for the objects; $\phi = (u_{ij}, (i,j) \in \Xi_q \times \Xi_q, q \in I_N)$ represents auxiliary variables in the quasi phi-function for scaled objects $\beta T_i(u_i)$ and $\beta T_j(u_j)$ (see [26]); $\Phi'^{T_iT_j}\left(u_i, u_j, u_{ij}, \beta\right)$ is the quasi-phi-function for $\beta T_i(u_i)$ and $\beta T_j(u_j)$; $\Phi^{T_i\Omega^*}\left(u_i, \beta\right)$ is the phi-function for $\beta T_i(u_i)$ and Ω^*;$\Phi'^{\widehat{\Lambda_q}\widehat{\Lambda_{gg}}}\left(z_q, z_g, \varphi_{qg}, \gamma_{qg}, \beta\right)$ is the quasi-phi-function for two clusters (see [26] for more details); $\sigma = 1 + 3n + \frac{\tau_q}{2}\sum_{q=1}^{N} m_q\left(m_q - 1\right) + N(N-1)$ is the number of the problem variables, τ_q is the number of auxiliary variables.

A feasible starting point $(u^0, \phi^0, \psi^0, \beta^0 = 0)$ is used to solve the problem (8.15)–(8.16). Here $\psi^0 = \left(\varphi^0_{qg}, \gamma^0_{qg}, q > g \in J_N\right)$ is obtained at Step 6; $v_i^0 = \left(x_i^0, y_i^0\right) \in C_q$ found at Step 5 and $\theta_i^0 \in [0, 2\pi]$ is a randomly generated rotation parameter of $\beta T_i(u_i)$ for $i \in \Xi_q$, $q \in I_N$; $\phi^0 = \left(u^0_{ij}, (i,j) \in \Xi_q \times \Xi_q, q \in I_N\right)$, a vector u^0_{ij} of auxiliary variables is obtained by simple geometrical subject to $\Phi'^{T_iT_j}\left(u_i^0, u_j^0, u_{ij}^0, \beta^0 = 0\right) \geq 0$.

Global solution to the problem (8.15)–(8.16), ($\beta^* = 1$), gives a solution to the original problem (8.5)–(8.6). It can be used as a starting point for the problem (8.5)–(8.6).

8.5 Computational Results

The instances from [26] were used to demonstrate the efficiency of our approach for packing clusters subject to balancing conditions. The computations were running on AMD FX(tm)-6100, 3.30 GHz computer, Programming Language C++, Windows 7. Free optimization IPOPT code https://projects.coin-or.org/Ipopt was used [30].

Six instances from [26] were considered for the equilibrium packing problem in a container Ω with $l = 15$, $w = 15$, $w_0 = 0$, $(x_e, y_e) = (7.5, 7.5)$ for the Instances 1–6 and $(x_e, y_e) = (6.5, 6.5)$ for the Instance 7, $\Delta x_e = \Delta y_e = 0.05$. Computational results are provided for packing clusters with and without balancing conditions. We run our program 100 times for each instance.

Instance 1. A collection of $n = 36$ circles from the Example 2 [26] is given. The circles are divided into clusters $\Lambda_q(z_q)$, $q = 1, 2, 3, 4 : m_1 = 10$, $m_2 = 10$, $m_3 = 8$, $m_4 = 8$;

$\{w_i, i = 1, \ldots, 36\}$= {2.235, 2.250, 2.253, 1.454, 1.018, 0.832, 0.664, 0.585, 0.386, 0.275, 0.113, 2.403, 2.350, 2.286, 1.535, 1.086, 0.893, 0.719, 0.628, 0.428, 0.310, 0.144, 2.576, 2.452, 2.320, 1.618, 1.156, 0.956, 0.776, 0.672, 0.472, 0.348, 0.180, 2.756, 2.557, 2.353}.

Our result: (a) without balancing conditions ρ^*= 0.410928, the CPU time is 137.796 s.; (b) with balancing conditions ρ^*= 0.424021, $\mu_1(v^*)$=0.008224, $\mu_2(v^*)$=0.049992, the CPU time is 275.919 s. The corresponding local optimal packings are provided in Fig. 8.1.

Instance 2. The collection of $n = 30$ ellipses from Example 1 in [26] is given. The collection of ellipses is divided into clusters $\Lambda_q(z_q)$, $q = 1, 2, 3 : m_1 = 10$, $m_2 = 10$, $m_3 = 10$;

$\{w_i, i = 1, \ldots, 30\}$= (3.000, 2.700, 2.400, 1.800, 1.300, 1.080, 0.880, 0.750, 0.540, 0.400, 0.210, 3.000, 2.700, 2.400, 1.800, 1.300, 1.080, 0.880, 0.750, 0.540, 0.400, 0.210, 3.000, 2.700, 2.400, 1.800, 1.300, 1.080, 0.880, 0.750).

Our result: (a) without balancing conditions ρ^*= 0.691521, the CPU time is 582.586 s; (b) with balancing conditions ρ^*= 0.688519, $\mu_1(v^*)$=0.019491, $\mu_2(v^*)$=0.037263, the CPU time is 1103.894 s. The corresponding local optimal packings are provided in Fig. 8.2.

Instance 3. A collection of $n = 36$ objects from Example 4 [26] is given. The objects are divided into clusters $\Lambda_q(z_q)$, $q = 1, 2, 3, 4$; $m_1 = 10$ circles, $m_2 = 10$ circles, $m_3 = 8$ ellipses, $m_4 = 8$ ellipses;

$\{w_i, i = 1, \ldots, 36\}$= (2.641, 2.481, 2.326, 1.626, 1.156, 0.951, 0.766, 0.660, 0.456, 0.331, 0.160, 2.641, 2.481, 2.326, 1.626, 1.156, 0.951, 0.766, 0.660, 0.456,

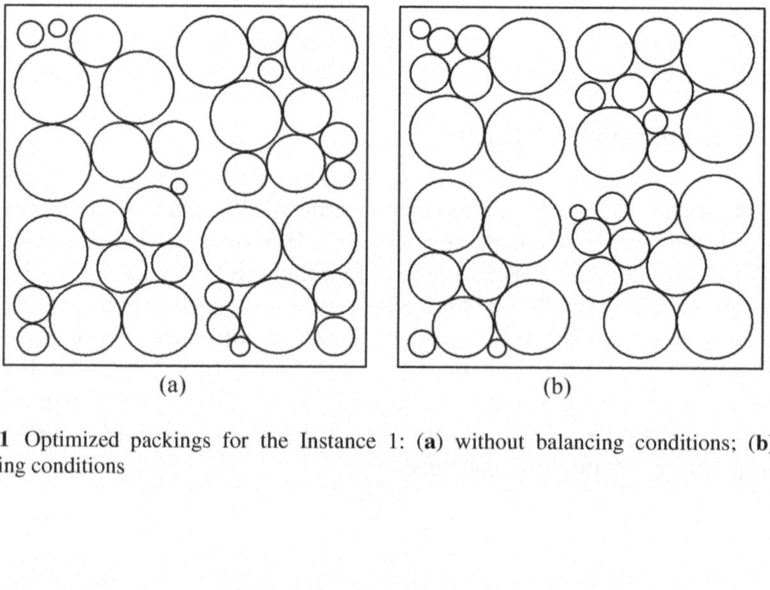

(a) (b)

Fig. 8.1 Optimized packings for the Instance 1: (**a**) without balancing conditions; (**b**) with balancing conditions

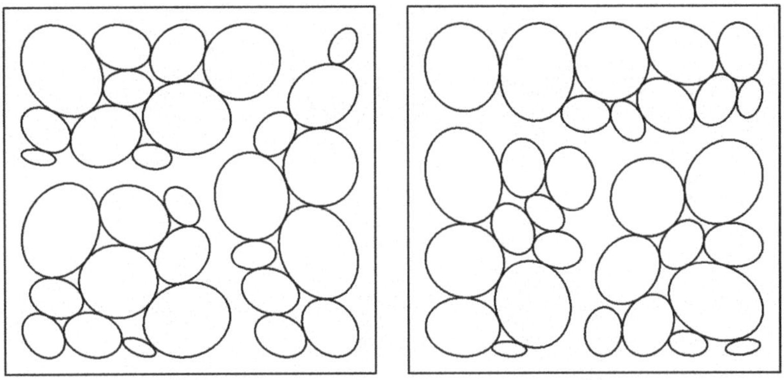

Fig. 8.2 Optimized packings for the Instance 1: (**a**) without balancing conditions; (**b**) with balancing conditions

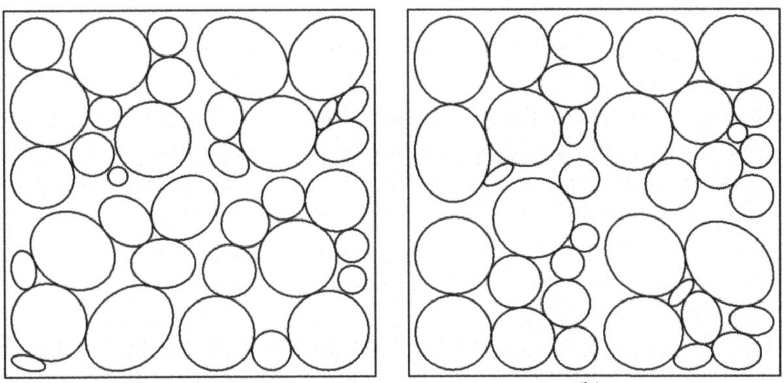

Fig. 8.3 Optimized packings for the Instance 3: (**a**) without balancing conditions; (**b**) with balancing conditions

0.400, 0.210, 3.000, 2.700, 2.400, 1.800, 1.300, 1.080, 0.880, 0.750, 0.540, 0.400, 0.210, 3.000, 2.700, 2.400).

Our result: (a) without balancing conditions $\rho^*= 0.272577$, $\mu_1(v^*)$=0.008624, $\mu_2(v^*)$=0.049986, the CPU time is 317.026 s; (b) with balancing conditions $\rho^*=$ 0.269150, the CPU time is 384.371 s. The corresponding local optimal packings are provided in Fig. 8.3.

Instance 4. A collection of $n = 36$ objects from Example 3 [26] is given. The ellipses is divided into clusters $\Lambda_q(z_q)$, $q = 1, 2, 3, 4$: $m_1 = 10$ circles, $m_2 = 10$ circles, $m_3 = 8$ ellipses, $m_4 = 8$ ellipses;

$\{w_i, i = 1, \ldots, 36\}$= {2.250, 2.250, 2.250, 1.440, 1.000, 0.810, 0.640, 0.563, 0.360, 0.250, 0.090, 2.250, 2.250, 2.250, 1.440, 1.000, 0.810, 0.640, 0.563, 0.360,

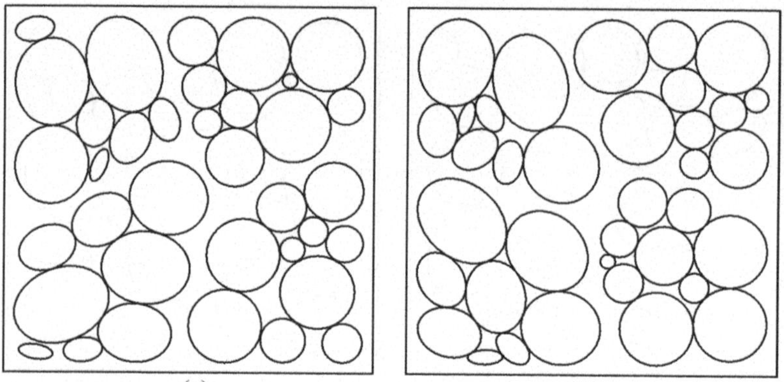

Fig. 8.4 Optimized packings for the Instance 3: (**a**) without balancing conditions; (**b**) with balancing conditions

0.400, 0.210, 3.000, 2.700, 2.400, 1.800, 1.300, 1.080, 0.880, 0.750, 0.540, 0.400, 0.210, 3.000, 2.700, 2.400}.

Our result: (a) without balancing conditions $\rho^*=$ 0.422594, the CPU time is 312.626 s; (b) with balancing conditions $\rho^*=$ 0.421740, $\mu_1(v^*)$=0.049982, $\mu_2(v^*)$=0.049982, the CPU time is 1925.068 s. The corresponding local optimal packings are provided in Fig. 8.4.

Instance 5. A collection of $n = 30$ objects from Example 6 [26] is given. The objects are divided into clusters $\Lambda_q(z_q)$, $q = 1, 2, 3 : m_1 = 10$ convex polygons, $m_2 = 10$ circles, $m_3 = 10$ ellipses;

$\{w_i, i = 1, \ldots, 30\}$= (4.000, 3.240, 2.890, 2.560, 1.960, 1.690, 1.440, 1.210, 1.000, 0.810, 0.113, 2.403, 2.350, 2.323, 2.289, 1.543, 1.513, 1.098, 0.904, 0.729, 0.400, 0.210, 3.000, 3.040, 2.700, 2.400, 1.800, 1.300, 1.080, 0.880). Our result: (a) without balancing conditions $\rho^*=$ 0.304910, the CPU time is 1564.550 s; (b) with balancing conditions $\rho^*=$ 0.269486, $\mu_1(v^*)$=0.050002, $\mu_2(v^*)$=0.032384, the CPU time is 1990.042 s. The corresponding local optimal packings are provided in Fig. 8.5.

Instance 6. An collection of $n = 36$ objects from Example 5 [26] is given. The objects are divided into clusters $\Lambda_q(z_q)$, $q = 1, 2, 3, 4 : m_1 = 10$ convex polygons, $m_2 = 10$ circles, $m_3 = 8$ ellipses, $m_4 = 8$ ellipses;

$\{w_i, i = 1, \ldots, 36\}$= (3.000, 2.700, 2.400, 1.800, 1.300, 1.080, 0.880, 0.750, 0.540, 0.400, 0.113, 2.403, 2.350, 2.286, 1.535, 1.086, 0.893, 0.719, 0.628, 0.428, 0.400, 0.210, 3.000, 2.700, 2.400, 1.800, 1.300, 1.080, 0.880, 0.750, 0.540, 0.400, 0.210, 3.000, 2.700, 2.400).

Our result: (a) without balancing conditions $\rho^*=$ 0.370691, the CPU time is 1543.848 s; (b) with balancing conditions $\rho^*=$ 0.344749, $\mu_1(v^*)$=0.050010, $\mu_2(v^*)$=0.049998, the CPU time is 1503.132 s. The corresponding local optimal packings are provided in Fig. 8.6.

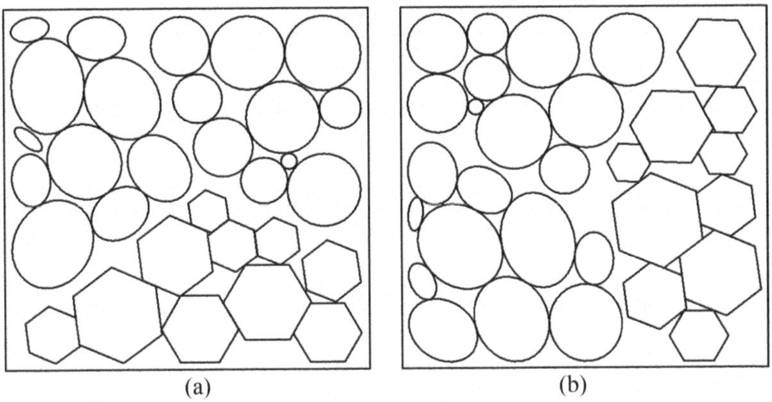

Fig. 8.5 Optimized packings for the Instance 5: (a) without balancing conditions; (b) with balancing conditions

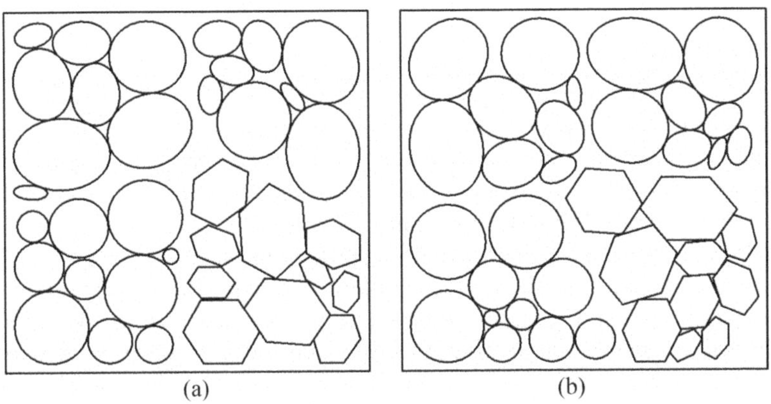

Fig. 8.6 Optimized packings for the Instance 6: (a) without balancing conditions; (b) with balancing conditions

Instance 7. A collection of $n = 30$ ellipses from Instance 1 is given with$(x_e, y_e) = (6.5, 6.5)$. Our result with balancing conditions is $\rho^* = 0.273300$, $\mu_1(v^*)$=0.050025, $\mu_2(v^*)$=0.050028. The CPU time is 542.01 s. The corresponding local optimal packing is provided in Figure 8.7a.

A collection of $n = 36$ circles from Instance 2 is given with $(x_e, y_e) = (6.5, 6.5)$. Our result with balancing conditions is $\rho^* = 0.014361$, $\mu_1(v^*)$=0.049988, $\mu_2(v^*)$=0.050003. The CPU time is 355.183 s. The corresponding local optimal packing is provided in Figure 8.7b.

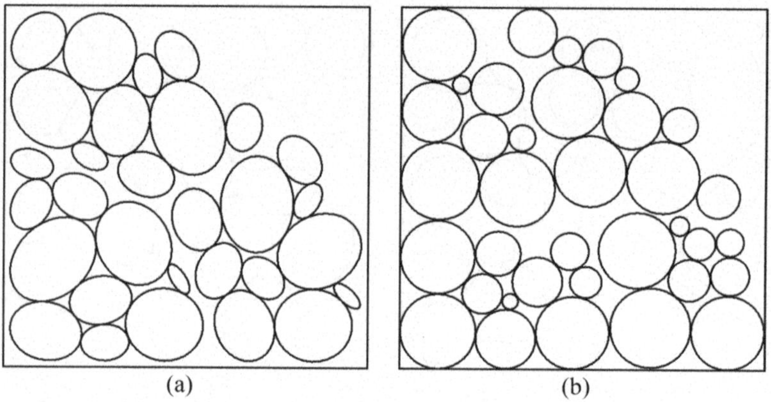

Fig. 8.7 Optimized packings with balancing conditions for the Instance 7: (**a**) for clusters of ellipses; (**b**) for clusters of circles

8.6 Conclusions

Packing clusters of convex objects is considered subject to balancing conditions. The latter is essential in container loading problems to guarantee equilibrium of the container during its movements. Packing is considered for a given rectangular container providing non-overlapping constraints for clusters as well as for objects within each cluster. Each cluster involves objects of the same shape.

A new mathematical model for the equilibrium packing clusters is presented and corresponding nonlinear continuous problem is formulated. To accelerate a local optimization procedure the algorithm to get good initial feasible arrangements is proposed. New benchmark instances are reported demonstrating the efficiency of the technique for equilibrium packing clusters composed of ellipses, circles, and convex polygons.

References

1. B. Chazelle, H. Edelsbrunner, L.J. Guibas, The complexity of cutting complexes. Discrete Comput. Geom. **4**(2), 139–181 (1989)
2. I. Litvinchev, L. Infante, L. Ozuna, Approximate packing: Integer programming models, valid inequalities and nesting, in *Optimized Packings and Their Applications*, ed. by G. Fasano, J. D. Pinter. Springer Optimization and Its Applications, vol. 105 (2015), pp. 117–135
3. J. Kallrath, S. Rebennack, Cutting ellipses from area-minimizing rectangles. J. Glob. Optim. **59**(2), 405–437 (2014)
4. Y. Stoyan, A. Pankratov, T. Romanova, Quasi-phi-functions and optimal packing of ellipses. J. Glob. Optim. **65**(2), 283–307 (2016)
5. E.G. Birgin, R.D. Lobato, J.M. Martinez, Packing ellipsoids by nonlinear optimization. J. Glob. Optim. **65**(4), 709–743 (2016)

6. A. Pankratov, T. Romanova, I. Litvinchev, Packing ellipses in an optimized rectangular container. Wirel. Netw. (2018). https://doi.org/10.1007/s11276-018-1890-1
7. F. Kampas, I. Castillo, J. Pintér, Optimized ellipse packings in regular polygons. Optim. Lett. **13**(7), 1583–1613 (2019)
8. A. Pankratov, T. Romanova, I. Litvinchev, Packing ellipses in an optimized convex polygon. J. Glob. Optim. **75**(2), 495–522 (2019)
9. P. Stetsyuk, T. Romanova, G. Scheithauer, On the global minimum in a balanced circular packing problem. Optim. Lett. **10**, 347–1360 (2016)
10. F. Zhanghua, H. Wenqi, Z. Lü, Iterated tabu search for the circular open dimension problem. Eur. J. Oper. Res. **225**(2), 236–243 (2013)
11. Y. Stoyan, G. Yaskov, Packing equal circles into a circle with circular prohibited areas. Int. J. Comput. Math. **89**(10), 355–1369 (2012)
12. H. Akeb, M. Hifi, S. Negre, An augmented beam search-based algorithm for the circular open dimension problem. Comput. Ind. Eng. **61(2**, 373–381 (2011)
13. R. Torres, J.A. Marmolejo, I. Litvinchev, Binary monkey algorithm for approximate packing non-congruent circles in a rectangular container. Wirel. Netw. (2018). https://doi.org/10.1007/s11276-018-1869-y
14. I. Litvinchev, L. Infante, L. Ozuna, Packing circular like objects in a rectangular container. J. Comput. Syst. Sci. Int. **54**(2), 259–267 (2015)
15. J. Kallrath, Cutting circles and polygons from area minimizing rectangles. J. Glob. Optim. **43**, 299–328 (2009)
16. J. Peralta, M. Andretta, J. Oliveira, Packing circles and irregular polygons using separation lines, in *Proceedings of the 7th International Conference on Operations Research and Enterprise Systems (ICORES 2018)*, (2018), pp. 71–77
17. H. Alt, F. Hurtado, Packing convex polygons into rectangular boxe, in *JCDCG'00 Revised Papers from the Japanese Conference on Discrete and Computational Geometry*, (Springer, London, 2000), pp. 67–80
18. H. Alt, M. Berg, C. Knauer, Approximating minimum-area rectangular and convex containers for packing convex polygons. J. Comput. Geom. **8**(1), 1–10 (2017)
19. D. Jones, A fully general, exact algorithm for nesting irregular shapes. J. Glob. Optim. **59**, 367–404 (2013)
20. F.M. Toledo, M.A. Carravilla, C. Ribeiro, J.F. Oliveira, A.M. Gomes, The dotted board model: a new MIP model for nesting irregular shapes. Int. J. Prod. Econ. **145**(2), 478–487 (2013)
21. Y. Stoyan, A. Pankratov, T. Romanova, Cutting and packing problems for irregular objects with continuous rotations: mathematical modeling and nonlinear optimization. J. Oper. Res. Soc. **67**(5), 786–800 (2016)
22. Y. Stoyan, A. Pankratov, T. Romanova, Placement problems for irregular objects: mathematical modeling, optimization and applications, in *Optimization Methods and Applications*, ed. by S. Butenko, P. M. Pardalos, V. Shylo. Springer Optimization and Its Applications, vol. 180 (2017), pp. 521–559
23. J. Bennell, G. Scheithauer, Y. Stoyan, T. Romanova, A. Pankratov, Optimal clustering of a pair of irregular objects. J. Glob. Optim. **61**(3), 497–524 (2015)
24. A.A.S. Leao, F.M.B. Toledo, J.F. Oliveira, M. Carravilla, R. Alvarez-Valdés, Irregular packing problems: a review of mathematical models. Eur. J. Oper. Res. **282**(3), 803–822 (2019). https://doi.org/10.1016/j.ejor.2019.04.045
25. A. Bortfeldt, G. Wäscher, Constraints in container loading – a state-of-the-art review. Eur. J. Oper. Res. **229**(1), 1–20 (2013)
26. T. Romanova, A. Pankratov, I. Litvinchev, Y. Pankratova, I. Urniaieva, Optimized packing clusters of objects in a rectangular container. Math. Probl. Eng. **2019**, Article ID 4136430, 12 pages (2019). https://doi.org/10.1155/2019/4136430
27. Y. Stoyan, T. Romanova, Mathematical models of placement optimization: two- and three-dimensional problems and applications, in *Modeling and Optimization in Space Engineering*, ed. by G. Fasano, J. D. Pinter. Springer Optimization and Its Applications, vol. 73 (2012), pp. 363–388

28. N. Chernov, Y. Stoyan, T. Romanova, A. Pankratov, Phi-functions for 2D objects formed by line segments and circular arcs. Adv. Oper. Res. (2012). https://doi.org/10.1155/2012/346358
29. T. Romanova, J. Bennell, Y. Stoyan, A. Pankratov, Packing of concave polyhedra with continuous rotations using nonlinear optimization. Eur. J. Oper. Res. **268**, 37–53 (2018)
30. A. Wachter, L.T. Biegler, On the implementation of an interior-point filter line-search algorithm for large-scale nonlinear programming. Math. Program. **106**(1), 25–57 (2006)

Part III
Simulation and Distributed Systems

Chapter 9
Didactic Tool for Teaching Election Algorithms in Distributed Systems

Araceli López-Reyes and Francisco de Asís López Fuentes

Contents

9.1 Introduction 111
9.2 Related Work 112
9.3 Chang's Algorithm 113
9.4 Analysis and Design 114
9.5 Implementation 118
9.6 Evaluation 120
9.7 Conclusions 123
References 123

9.1 Introduction

The digital revolution has had a high impact on various sectors of society, which demand more competitive students as in both theoretical and practical aspects. To improve the teaching-learning processes many researchers have identified that ICT integration into the education system can improve its quality. In this article we present a didactic tool to facilitate the learning of the algorithms distributed to the students of the courses of distributed systems at the undergraduate level. Our main motivation to create this tool was the need to be able to transmit new knowledge to different people. Our teaching tool aims to strengthen the teaching-learning process by developing a graphical user interface with the necessary requirements to provide an experience that allows learning and developing knowledge of the election algorithms in distributed computing. Specifically, in this paper the operation of the Chang ring algorithm is presented. Our objectives are (a) design of visualization interface for teaching material, (b) design and implement of visualization of

A. López-Reyes · F. de Asís López Fuentes (✉)
Department of Information Technology, Universidad Autónoma Metropolitana – Cuajimalpa, Mexico City, Mexico
e-mail: flopez@correo.cua.uam.mx

P. Vasant et al. (eds.), *Data Analysis and Optimization for Engineering and Computing Problems*, EAI/Springer Innovations in Communication and Computing,
https://doi.org/10.1007/978-3-030-48149-0_9

election algorithms in distributed computing, in this case the Chang ring, (c) use of animations to transmit knowledge in such a way that any user finds it understandable.

The rest of this work has the following organization. In Sect. 9.2, information about teaching tools related to distributed systems is presented. Section 9.3 introduces Chang's algorithm. An analysis and design of the didactic tool based on this algorithm is described in Sect. 9.4. The implementation of our teaching tool is described in Sect. 9.5. Section 9.6 presents the tests and evaluations performed to our tool. The article concludes in Sect. 9.7.

9.2 Related Work

Many researchers around the world are integrating information and communication technologies (ICT) in education. Although different teaching tools have been developed for teaching computer courses, there are still few teaching tools for teaching distributed systems. In this section we describe some example where ICT are integrated in the teaching-learning process.

A didactic tool related to a virtualized game is presented in [1]. In this work the authors believe that there is a potential in the digital games to explore in different areas of education. Scientific associations in the USA suggest giving priority to these games in the teaching to develop new skills as planning, strategy, or adaptation capacity in the students [1]. The objective of this tool is to present students with a non-trivial game and provide a game environment that presents different cases of failures to solve. This scenario forces students to have to investigate and understand new topics where they have not had previous experience. Thus, they gain experience to evaluate a pool of different causes to solve a problem. With this tool, students are expected to have experience in a scenario that teaches them and that allows them to push their ability to scale to the limit and know the possible situations where a system can go from bad to worse and that allows them to detect in time and form the root cause of a problem, thus promoting their knowledge both in practice and in theory. During the creation of the tool described above, the authors had the need to create games that allow people to promote analytical skills that in the professional world are necessary. The main objective of this tool as well as the tool that we present in this work is to be able to provide the student with the experience of being able to investigate and understand complex distributed systems. In turn, it was sought that students could see different types of scenarios where a distributed system can go wrong or different cases where it is necessary to analyze and find a root solution. In this project students showed interest in the application and this was funny and addictive for them. It was observed that the students worked together and were able to solve the problems. The students were able to learn how the system works in an outstanding way, which caused them to solve two problems in the system in 2 h, with the basic knowledge that was acquired. An interactive tool to learn the basics of the sensor network is presented in [2]. In this work the authors take for granted the need to be able to transmit basic knowledge to future generations. On this occasion the main objective for the development of this tool was that the students could freely explore the routing protocols. Wireless sensor networks consist of intelligent autonomous sensor nodes, which can measure characteristics of their environment, such as temperature, pressure, noise level, acceleration, etc. The WSN (wireless sensor network) has a wide range of applications such as climate monitoring, tropical flood prevention, or forest fire detection. During the elaboration of this project, the concern of the useful life of the batteries when sending status messages was explained, being that it is sought to be able to extend this useful life without compromising the status messages that alert about possible events. Once this project was completed, it was found that the students are more aware of the many particularities of the routing protocols, all this while they are playing and acquiring knowledge that allows exploring various solutions by acquiring such knowledge more efficiently and easily.

Recently, other strategies or teaching tools that integrate ICT have been presented in [8–11]. Authors in [9] explore how the use of video in the vocational teacher education can experiment the professional learning. Authors found that video can promote and support student's professional growth, because they learn new ways to work with the information, and a best understand between the theory and practice in their professional activity. However, a teaching based on video has limitations when variable conditions are changing. In [8], authors present different methods as a pedagogical important way to change the teaching and learning for specialized courses in computation. A model and framework are proposed in order to give to students an opportunity to manipulate their own examples along with the theory lectures in computer networks. In WiFiSiM [10] the authors have created a didactic tool to improve the teaching and learning of computer networks. This didactic tool is mainly focused to simulate the behavior of IEEE 802.11 network protocols. This software allows users to design and model wireless networks in a realist way, because realism of the simulation is one of its most important characteristics. A highly configurable graphic user interface is used to improve the interaction with the students. Finally, authors in [11] present a case about education with mobile device in order to study the effect of this strategy in the teaching and learning processes. Advantages and disadvantages of the mobile learning based on analysis of individual aspects and contents are reviewed.

9.3 Chang's Algorithm

Chang's algorithm is an algorithm of choice that is based on the principle of selective extinction. This algorithm considers that each process has a channel with the following process in the ring and the process that starts the algorithm is marked as a participant and sends its identifier in a message of choice to its neighbor. Messages circulate clockwise. This algorithm should be used when:

- The processes are physically or logically arranged in a ring.
- The total number of processes (n) is unknown.

- Each process communicates with its neighbor (left or right).

The process to be carried out [3, 4] is detailed below:

1. Initially all processes are "non-participants."
2. Any process P decides to start an election at any time.
3. This process P is put in a "participant" state and sends a message of choice M to its neighbor.
4. The message M sent contains the ID (identifier) of the process that initiated the election.
5. When the neighbor receives the choice message M, it establishes its status as a participant and checks the message ID.
6. If it is greater than its own ID, then send it directly to its neighbor.
7. If its ID is greater than the received ID, then place it in the M message and send it to its neighbor.
8. This is how message M is circulated successively until it reaches a Pn process that verifies that the received ID is its own. That indicates that only the largest ID has survived, which is that of the Pn process.
9. Then, this Pn process is the coordinator and notifies its neighbor. When a process receives a coordinator message, it must set its status as "non-participant" and send the message to its neighbor.
10. When the coordinator message returns to the process that issued it (the coordinator), then all the processes know who the coordinator is and everyone is in a "non-participating" state.

If two processes initiate a choice at the same time and election messages are sent, a process in a "participant" state must verify the ID of the process that sends the election message. If it is less than its own, the message is discarded. Thus, all messages of choice will be extinguished, except the one with the highest ID.

9.4 Analysis and Design

This section explains the analysis and design that was carried out prior to the implementation of the Chang ring tool. The analysis and design that was carried out to have the final tool considers that this tool will have a didactic use. Several tests were performed so that it was user friendly, as well as its use was intuitive.

Requirements Analysis

Before the design and implementation of the Chang ring tool, the functions that it would do and what the user could visualize when using the tool were established. In the same way a design was designed to do it in a didactic and friendly way. Chang ring tool allows the user the possibility to learn an election algorithm through a web page, which is accessible to all users who wish to learn the use and operation of it.

Users through the interface will be able to visualize how Chang's algorithm works. Each time an example is executed, users can see step by step how the algorithm is executed. The user is given the option on the initial page to select the number of nodes with which they want to display the example. This can select from 3 to 20 nodes. At the beginning of the demonstration, the moment in which the coordinating node is located is shown and when the other nodes are notified of this.

Tool Structure

The teaching tool consists of a dynamic web interface that allows the user to initially see a short explanation of what the Chang ring algorithm consists of. The user has the possibility through a drop-down list to select the number of nodes with which he wishes to display the example. After randomly, values are assigned to each of the nodes (without repeating) before beginning with the explanation of how the algorithm works. The user can visualize step by step how the algorithm works and compares the value of each node until finding the node that will be the coordinator. The languages to use in this project are HTML, CSS, and JavaScript [5–7]. These languages are commonly used for the design and creation of dynamic web pages. Figure 9.1 shows the navigation map of the web page and indicates the parts that make up the tool.

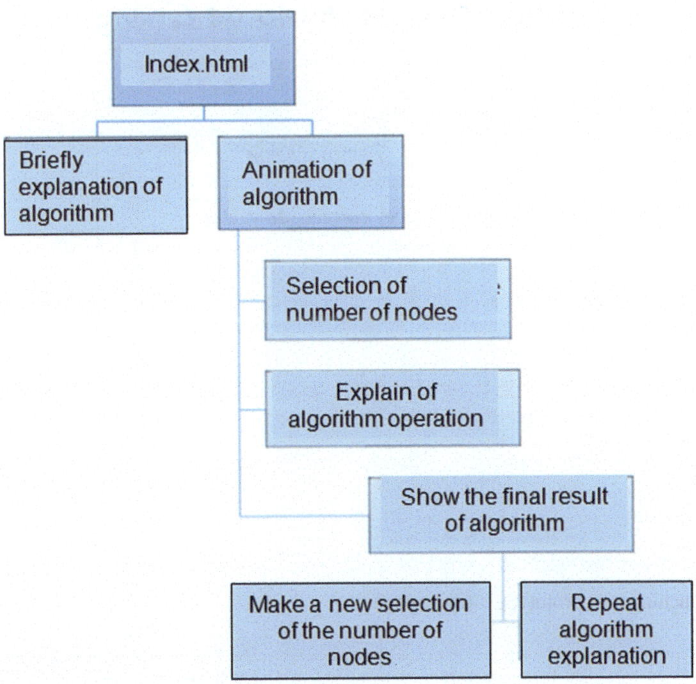

Fig. 9.1 Web tool navigation map

Interface Prototypes

For the design of the didactic tool, several designs were made before having the final design. Figure 9.2 shows the prototype of the teaching tool which allowed us to visualize more clearly how it would begin with the design of the tool. This figure shows the main functions of the main page.

Figure 9.3 shows the design to be displayed once the user selects the number of nodes for the demonstration. This figure shows when the algorithm found the value of the node that will be the coordinator. It also shows the "Notify who is the coordinator" button that the user must select to continue the process. Figure 9.4 shows the design once the nodes were notified of who the new coordinator is. In this case the user has the option to see the demonstration again in case he/she wants to review a step in detail or to select another demonstration. That is, the user can choose a different number of nodes to create, but the identifiers assigned to each of them will be different.

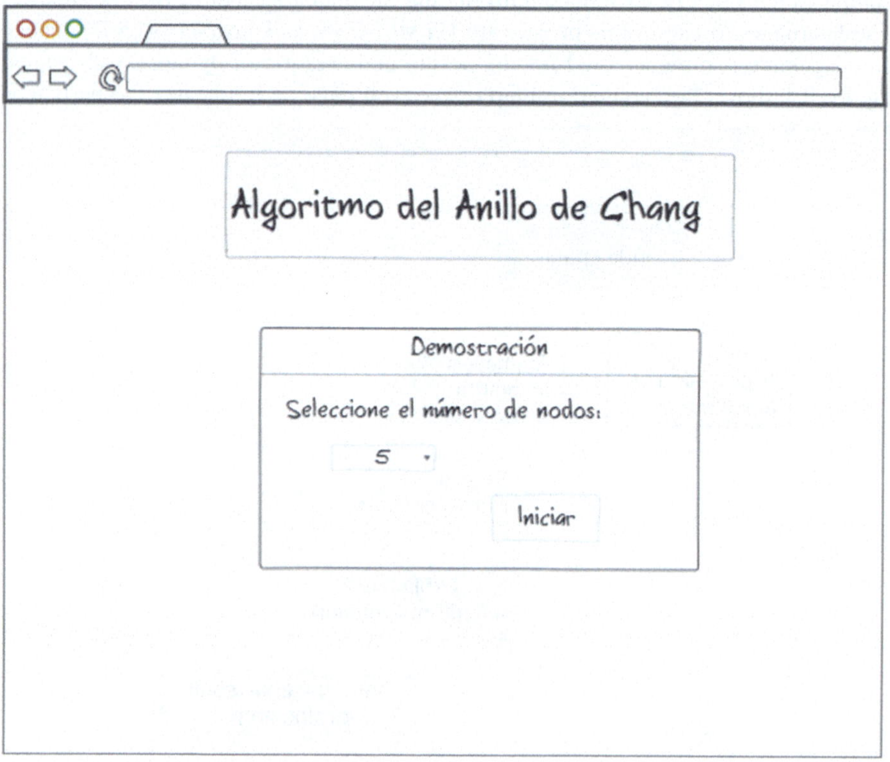

Fig. 9.2 Teaching tool prototype—home page

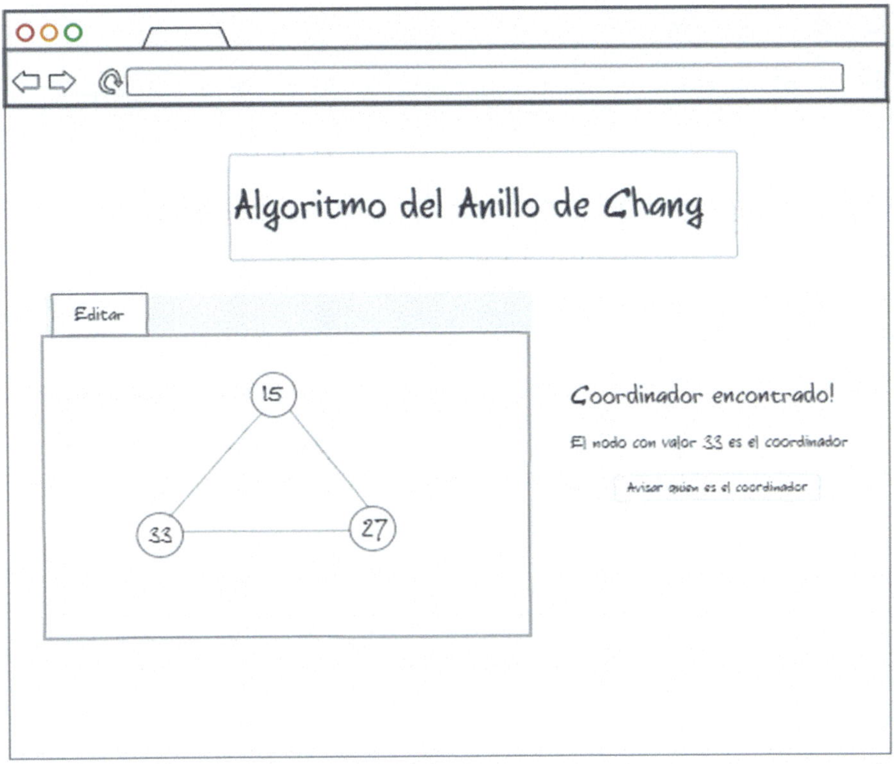

Fig. 9.3 Interface prototype showing the created nodes

Object Repository Design

The files of our didactic tool are stored strategically according to their extension and function in it, this to facilitate its handling and manipulation. The structure of the tool is as explained below. The project directory is created as the main directory, and all the files are required for its operation reside. There is a file called index.xml, which is the index of the project. That is, this is the main page through which the user can access the main functions of the tool. A CSS subdirectory is created to store the style sheets created and used for the use of the Chang ring tool, this in order to make the tool more attractive to the user. There is subdirectory images where the images that were used on the website are stored, this to make the tool highly intuitive and easy-to-interpret. Finally, a JS subdirectory is created to store the JavaScript scripts as well as the libraries used in the Chang ring tool.

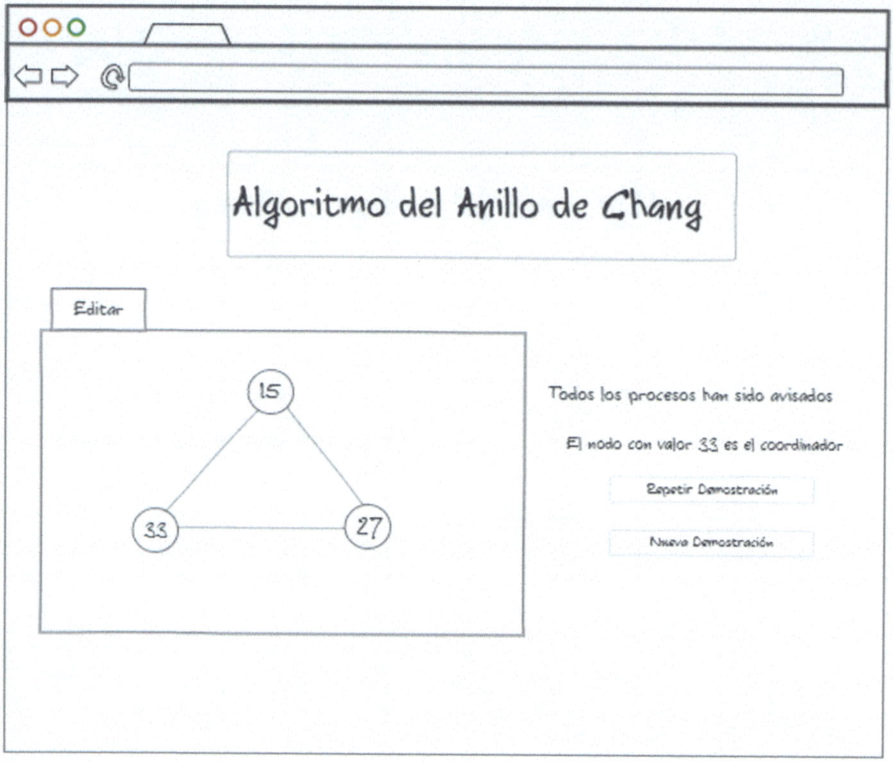

Fig. 9.4 Interface prototype after nodes have been notified of the new coordinator

9.5 Implementation

To implement the Chang ring tool, HTML, CSS, and JavaScript technologies were used. It began by implementing the structure of the directories that make up the tool (see Fig. 9.5). These are the project directories where subdirectories are organized for the files used by the tool.

The CSS subdirectory contains all the style sheets needed to format or present the web interface. The JS subdirectory contains all the JavaScript files that are necessary for the proper functioning of the system. The images subdirectory contains all the images that are used to give a better presentation to the web interface. The CSS subdirectory contains the cascading style sheets and the files that define the structure of the teaching tool. By having them sorted in this way we prevent the files from being very heavy, achieving a simpler and sharper way of working. The main libraries used for the design and layout of the didactic tool were Bootstrap and Vis. Bootstrap.css is a configuration of global cascading style sheets, as well as fundamentally designed and improved HTML elements through the use of extensible classes and an advanced grid system. The instructions.js file contains the

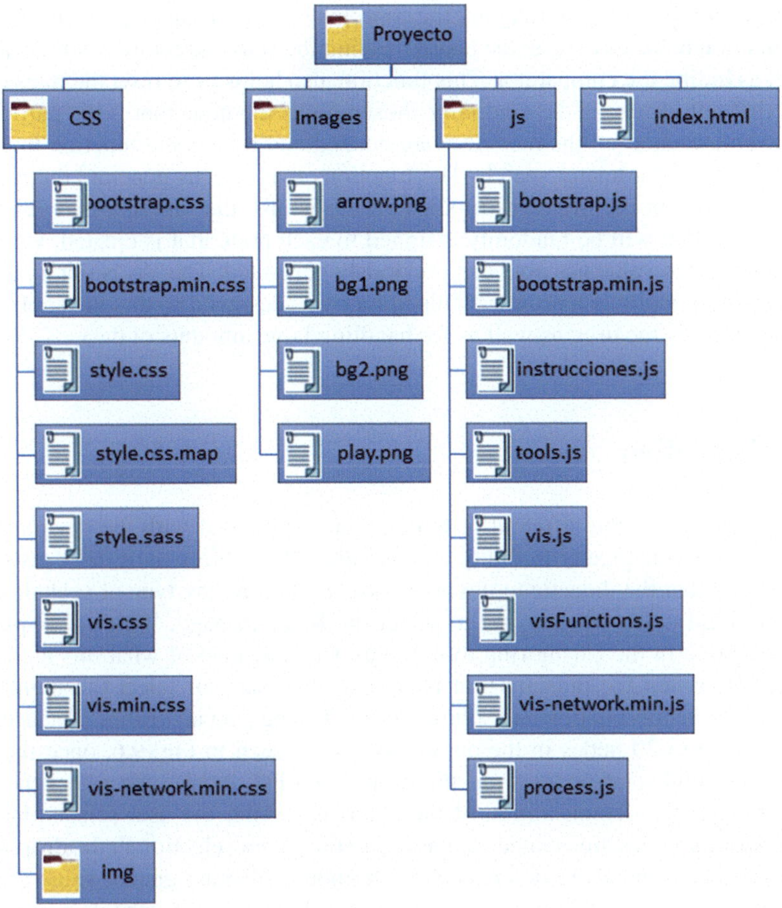

Fig. 9.5 File structure

instructions that are displayed on the screen, while the animation of the Chang ring algorithm is displayed. In process.js there is the function that will create the Chang ring, it is conditional that in case there is already a previous ring, delete it and create another one.

In this file is declared the array that avoids duplication of numbers, the variable that stores the maximum process identifier, the node identifier which is used to access it. Other functions found in the process.js file are createConnections, notifyNodes, and startAgain. With createConnections the lines connecting the nodes are created, first it is verified that it is not the final node and in case it is, the content is deleted, it shows the second step and with the preiteration function the user will be explained 3 instructions that will do before the iteration. Once the coordinating node was found, the notifyNodes function is responsible for informing the other nodes of its id. After notifying the nodes, print a message on the screen with the caption:

"End of example." With startAgain, the coordinator search can be performed again. This function helps us to reset the boards, hiding the start and remove buttons again, as well as hiding the comparator. This function also helps us to reset the flags of the nodes that tell us who is the randomly chosen node, the node that is the participant and the coordinator, all the messages are deleted and it calls the function in which it explains the coordinator search. The functions that help us generate the random color are saved in the file named tool.js. There is also the function that generates the numbers that will be randomly assigned to each node that is created, verifying that each created value has not been previously generated. The Vis.js library allows viewing dynamically in a browser. This library was designed so that the application is easily used by the user, as well as for handling large amounts of data.

9.6 Evaluation

This section shows the different tests that were performed with the teaching tool in order to provide greater usability and handling. The objective of these tests is to demonstrate that the didactic tool works correctly before any type of selection that the user could make when trying to select on the main page. With the display of the description in the Chang ring tool, the user is informed of what this algorithm consists of. When the didactic tool is started, the user can select the number of nodes for the representation and solution of the Chang ring algorithm. The user can select from 3 to 20 nodes in the demo. As can be seen in Fig. 9.6, once the user selects the number of nodes from the drop-down list, he/she can select the play button to start the demonstration of the algorithm. After the user selects the play button, several nodes are created, according to the user's selection in the drop-down list of number of nodes to be created. Each node is also assigned a value that we will call an indicator. This scenario is shown in Fig. 9.7. It is randomly indicated which node is the one that will start the process, and the user must select the "search for coordinator" button to continue the demonstration. The initial node compares its identifier (ID) with the node next to it, if the node has a larger identifier it becomes the new coordinating node. This process is done with all nodes. While each node is being compared, it is indicated with a different color in which node the route goes.

Fig. 9.6 Setting of nodes number (process)

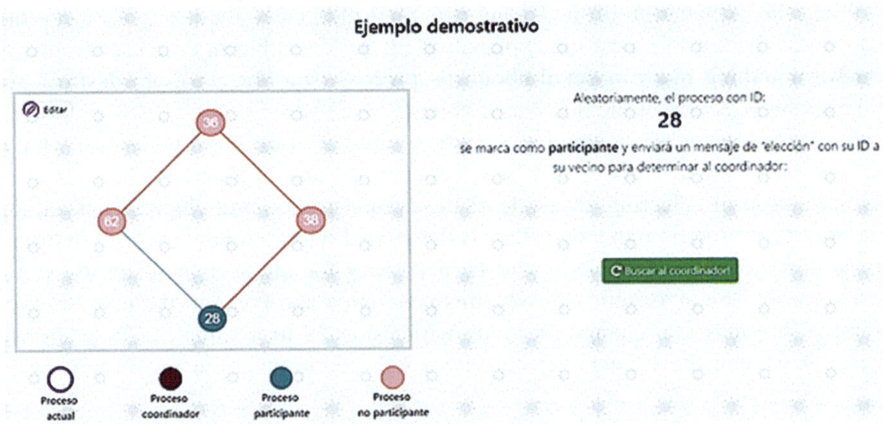

Fig. 9.7 Visualization of nodes with assigned identifier

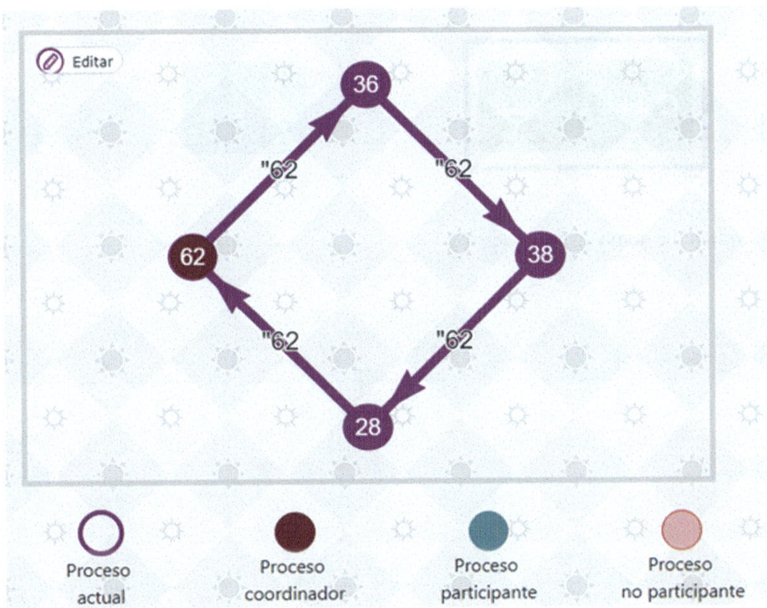

Fig. 9.8 Notification of new coordinator

Once you have finished comparing the identifier with all the nodes, it is determined who the coordinating node is, and the user must select the continue button to notify the other nodes who is the new coordinator. This is indicated with a different color to the node that has been notified. For example, in Fig. 9.8 node 62 indicates to the other nodes that it is the new coordinator.

When the explanation of the Chang ring algorithm ends, the user can select the "start over" button to repeat the explanation of the algorithm in case user wants to review something more in detail about the process (see Fig. 9.9), or "destroy" to make a new selection of the number of nodes.

A user can make a new simulation using a number of different nodes, for which the previously created nodes and their connections are eliminated. The user also has the option of selecting any node of the demonstration and changing its value, this in order to see how the algorithm is developed with that new change. In order to see how this setting works it will be necessary for the user to select the view again option. The algorithm will take this new value and perform the process. For example, the user can change the node with value 28 to a value of 72 as shown in Fig. 9.10, and the Chang ring algorithm will be simulated again. In this case,

Fig. 9.9 Watch the demo again or destroy

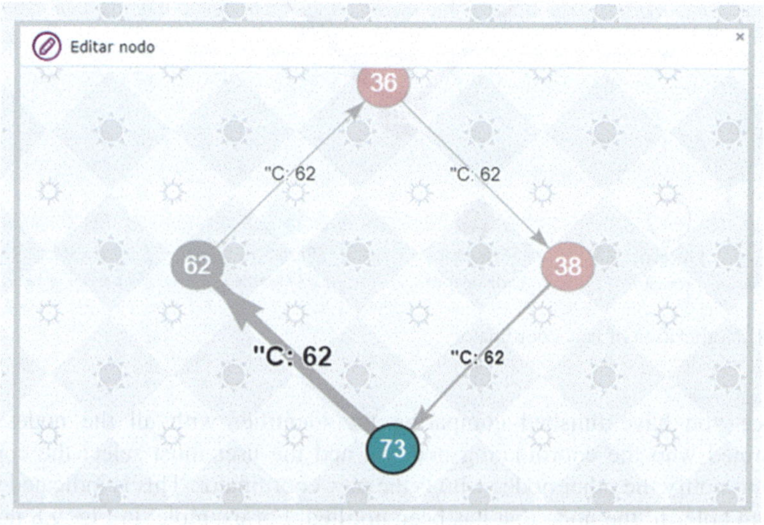

Fig. 9.10 A demonstration with different value for a new node

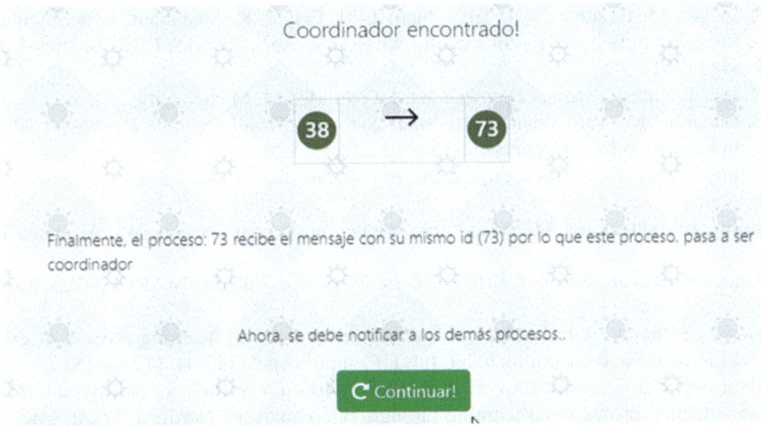

Fig. 9.11 A new coordinator is found

Fig. 9.11 shows in the new simulation the new coordinator found, which is the node with the value 72 entered by the user. This new coordinator must notify all user nodes as shown for the node with a value of 62 in Fig. 9.8.

9.7 Conclusions

A didactic tool for teaching distributed algorithms is presented in this work. Although there are similar systems that could teach knowledge effectively, we could see that tools such as the one designed in this work do not exist as such. However, if there are teaching systems such as the video games, which use different methods to achieve the same goal. Although a game is a system that could achieve similar and even more attractive results for the general public, we believe that there should be tools that facilitate acquiring knowledge before reaching this point and serve as a proof of acquired knowledge. As work in the future, an interactive game can be implemented to explore knowledge in a test environment that takes the student from a non-complex situation to a situation where the algorithm fails, under these conditions the students could thoroughly understand the limitations and improvements they could make to their knowledge. We also plan to add other distributed algorithms to our proposed scheme.

References

1. J. Wein et al., Virtualized games for teaching about distributed systems, in *Proceedings of the ACM Technical Symposium on Computer Science Education*, (ACM, New York, 2009), pp. 1–13

2. M.-P. Uwase, M. Bezunartea, J. Tiberghien, J.-M. Dricot, K. Steenhaut, Experimental comparison of radio duty cycling protocols for wireless sensor networks. IEEE Sensors J. **17**(19), 6474–6482 (2017)
3. F.A. López Fuentes, *Sistemas Distribuidos*, 1st edn. (UAM, Mexico City, 2015)
4. A. Tanenbaum, M. Van Steen, *Distributed Systems: Principles and Paradigms*, 2nd edn. (Pearson Education Inc., Upper Saddle River, 2008)
5. J. Eguiluz, Introducción a JavaScript, https://www.jesusda.com/docs/ebooks. Accessed 18 Aug 2019
6. J. Eguiluz, Introducción a XHTML, https://www.jesusda.com/docs/ebooks. Accessed 18 Aug 2019
7. J.D. Gauchat, *El gran libro de HTML5, CSS3 y Javascript*, 1st edn. (MARCOMBO, Barcelona, 2012)
8. M. Vinay, S. Rassak, A technological framework for teaching-learning process of computer networks to increase the learning habit. Int. J. Comput. Appl. **117**(4), 112 (2015)
9. D. Alvunger, V. Johansson, Exploring recontextualization of didactic ability and vocational teacher students' professional learning through video analysis. Nordic J. Vocat. Educ. Train. **8**(3), 36–56 (2018)
10. T.J. Mateo Sanguino, C. Serrano Lopez, F.A. Marquez Hernandez, WiFiSiM: an educational tool for the study and design of wireless networks. IEEE Trans. Educ. **56**(2), 149–155 (2013)
11. Y.T. Sung, K.E. Chang, T.C. Liu, The effects of integrating mobile devices with teaching and learning on students' learning performance: a meta-analysis and research synthesis. Comput. Educ. **94**, 252–275 (2016)

Chapter 10
Literature Review: Evaluation of the Feasibility of Implementing Industry 4.0 Technologies in the Intralogistic Processes of the Logistics Operators of the Department of the Atlantic, a Look Towards the Continuous Improvement of Organizational Efficiency

Jania Astrid Saucedo Martínez and Carlos Regalao Noriega

Contents

10.1 Introduction 125
10.2 Methodology 128
10.3 Discussion and Analysis of Information 134
10.4 Conclusion and Future Work 138
References 139

10.1 Introduction

To assess the feasibility of industry 4.0 technologies in the intralogistic processes of logistics operators in Colombia's Atlantic department, literature review should be initiated. Preliminary findings note that the topic has been developed immersed in the logistics area and topics related to inbound logistics management, technological

J. A. Saucedo Martínez
Universidad Autónoma de Nuevo León, San Nicolás de los Garza, Mexico

C. R. Noriega (✉)
Faculty of Mechanical and Electrical Engineering, Universidad Autónoma de Nuevo León, San Nicolás de los Garza, Nuevo León, Mexico

Faculty Engineering, Universidad Simón Bolívar, Barranquilla, Atlántico, Colombia
e-mail: cregalao@unisimonbolivar.edu.co

P. Vasant et al. (eds.), *Data Analysis and Optimization for Engineering and Computing Problems*, EAI/Springer Innovations in Communication and Computing,
https://doi.org/10.1007/978-3-030-48149-0_10

storage management, and automation in specific aspects such as barcodes, that have become one of the most important tools of companies when it comes to capturing data and having control of their products, so much so that 95% of 23 million Colombian companies code their articles with this modality, to have a unique identification [1]. Therefore, to meet the demands of the environment, and the international markets with which Colombia today has free trade agreements among the most important the Pacific Alliance it becomes necessary that the processes in the logistics operators adopt new tools to promote the smooth improvement of their business efficiency. However, these needs in the intralogistics area day by day will require more complex solutions than simply the use of barcodes and can be developed depending on the proposed simulation model for intralogistics activities, covering the processes of storage, loading and unloading of raw materials and the finished product of logistics operators, where these three procedures add up to almost 30% of the logistical costs of organizations in Colombia, which, according to the National Planning Department, account for approximately 15% of a company's sales [2]. For Gaviria [3] "From the year called 'Colombia is logistics' the May-res costs of this activity are in transport 37%; 20% storage; 17% supplier management and management; Sales order processing 10%; index and inventory replenishment 9% and reverse logistics 7%." This establishes as a need for logistics operators in the Atlantic department to find a tool that allows them to improve decision-making in terms of establishing the best technology for Industry 4.0 to use that they can use increase business efficiency in organizations.

The need to find a solution to the problems and challenges that the intralogistics area presents us together with the factors and variables coming from the environment and dynamics of organizations leads us to review in the literature the progress that has been developed about this subject. However, it is noteworthy that for Latin America and in particular attention Colombia are among the most lagging regions, for example, in terms of the Logistics Performance Index [4]. To improve these figures it depends not only exclusively on the governments of the region but also on the combination of a joint intralogistics effort in organizations, which through technological tools provided by Industry 4.0 can optimize productivity and improve business efficiency [5]; however, the Logistics Performance Index established by the World Bank, the country does not advance in logistics, on the contrary, it gives up space in the ranking of the region, going from position 10–13 between 2016 and 2017 [6]. It is therefore stated that the studies in this topic are in debt and progress in this area is mainly due to the improvement of the supply chain, therefore the topic is not developed independently, nor is it given the importance really deserves as Arsam sets it [7] where it defines that "intralogistic activities the most important thing is that, if it is properly optimized, it can greatly improve a company's service". However, it is one of the essential elements on which the so-called Storage Loading operates, a concept referred to in increasing the effectiveness of an organization by using intralogistics and the use of industry 4.0 technologies, see Fig. 10.1., the who are taking the world [8], these technologies called 4.0 are not alien to the field of logistics and in specifics to intralogistics activities, their incorporation begins to develop to the point already evident in the literature sketches of the

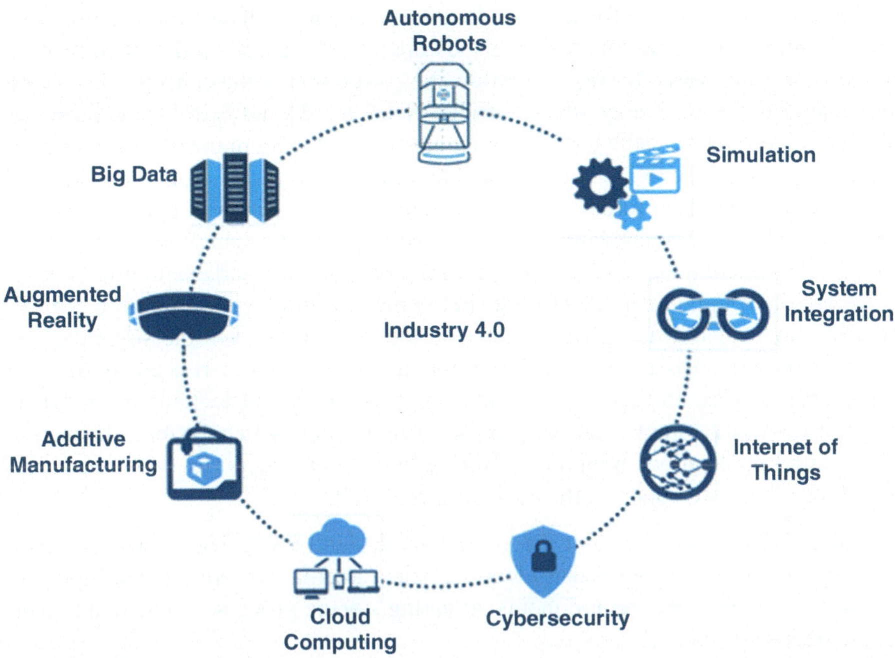

Fig. 10.1 Technologies and tools of industry 4.0

intralogistics concept 4.0 where J. Schuhmaher [9] defines it as "The control of future-oriented intralogistic systems," a large number of the flow control systems of supplies currently managed in the industry depend on centralized tools and, the emergence of new, effective and increasingly more sophisticated information and communications technologies [10]. The prerequisite for intralogistics 4.0 is data transparency [11]. Intelligent optimization allows, for example, that all transport orders of a logistics operator can be efficiently assigned to drivers and means of transport [12].

With the study of intralogistics framed in industry 4.0, the study of the Storage Loading concept is developed, framed in this concept assumed by Maslarić [13] who defines this correlation as the formation of global networks that will include installations of products and storage in the form of Cyberphysical Systems that would interact with each other in a non-dependent manner, are generated and controlled, which together with the previous concepts allow a continuous improvement in decision-making for the growth of business efficiency [14]. Among the authors studied, it is the one that resembles an ideal concept to the solution proposed in the present investigation and meets some of the characteristics of the solution, however, it does not contemplate the integration of the components of the intralogistics as a whole according to the principle of Bertalanffi, rather, he understands them as independent fragments of each other and with very minimal or isolated relations between them.

An important practice for the simulation model and evaluates the feasibility of implementing 4.0 technologies in logistics operators, focuses on the identification of the factors and variables that will allow the correlation of the concepts developed and analyzed in this literary review article. Neeraja [15] and Ltifi [16] determined the steps to be taken to address the challenges we must meet in the medium term, as a strategy to adapt the logistics means and identify the necessary tools. Logistics and intralogistics must therefore be studied and delimited as the issues of great relevance and necessity in the organizations of this century [17]. The central objective of the study is ingrained in the selection of opportunities, schemes, and exemplifications in the adoption of the numerous 4.0 tools in the field of intralogistics through critical, creative, and transformative thinking, from the evaluation scientific supplies, it is consistently expected to be a review item in future research related to selected categories such as: storage load, intralogistics, industry and technology 4.0. It is of great relevance to note that the process of reviewing and analyzing information did not find publications with an analysis similar to the categories developed. The definitions that correspond to the concepts are as follows:

- Kaiser [18], establishes that the value how design (VSD) The approach uses a morphology for the VSD and identifies when alternatives are studied based on specific supply chain requirements, affecting storage processes, load and unload the finished product.
- Storage Loading is constituted as a more applied storage tool, its most relevant feature consisting of a corridor that satisfies two adjacent rows of racks, has several corridors to operate throughout the storage path [19].
- A holistic system belonging to information technology activities, in the areas of people, machines, and tools [20], which warns in the flow of goods, services, and data in a controlled form [21], through the supply chain, with high-grade operations in autonomy [22] and great capacity to issue and deliver useful information to decision-making [23, 24].

This study is outlined as follows: Sect. 10.2 explains the methodology for the analysis of the data obtained, Sect. 10.3 establishes the important elements that are involved in accordance with the study and analysis of the data, and ends with the results of Sect. 10.4 that visualizes findings and highlights areas of opportunity for other research in the future.

10.2 Methodology

Methodological analysis in research development is established in the paradigm of qualitative analysis [25], the study includes four sections for greater understanding of the accuracy of the various sections; therefore, the methodology based on qualitative data analysis is used. Table 10.1 shows in detail each section and its characteristics.

Table 10.1 Methodology

Research sections [26]	Features	Activity
Data compilation	It includes many articles related to review categories	Collect research-based articles in various areas that delimit logistics operators
Sampling planning	This strategy provides for discriminatory sampling [27] and descriptive analysis	Reviewing of the literature developed covered the period from 2014 to 2019
Choice of categories	Maximize opportunities to verify constructed arguments, as well as relationships between categories	Visualize the parameters under investigation within the technology and operations that have developed the themes of the categories
Material evaluation	This phase was carried out with the compilation of all the material in an organized way	Recognize trends and scientific and technological progress in industry and 4.0 technologies based on categories

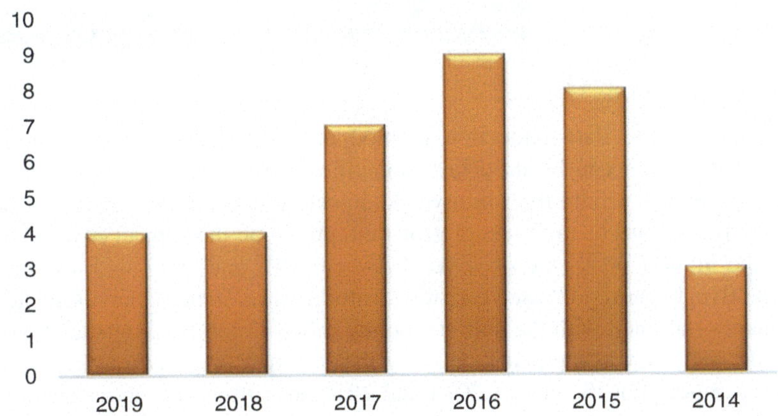

Fig. 10.2 Distribution of scientific articles per year

Method of Data Compilation

The data used for the adequacy and analysis of this article includes the time lapse between March 2014 and May 2019, see Fig. 10.2 considering the articles published in magazines, congresses, and seminars synthesis, in addition, accepted articles and in the publication process are on the web, so as not to allow publication of biased content. The research was conducted between April and July 2019. The relevant elements used in the search are delimited by words such as: "Industry 4.0," "Intralogistics," "4.0 technologies," "Storage Loading," "supply chain." These words are a total of 57 scientific articles to check see Fig. 10.3.

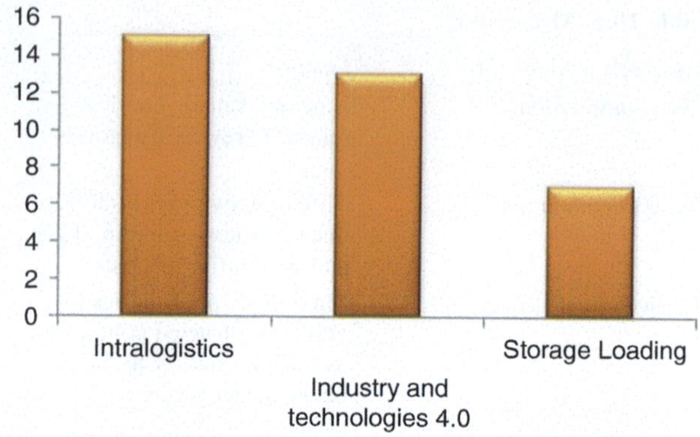

Fig. 10.3 Distribution of scientific articles by topic

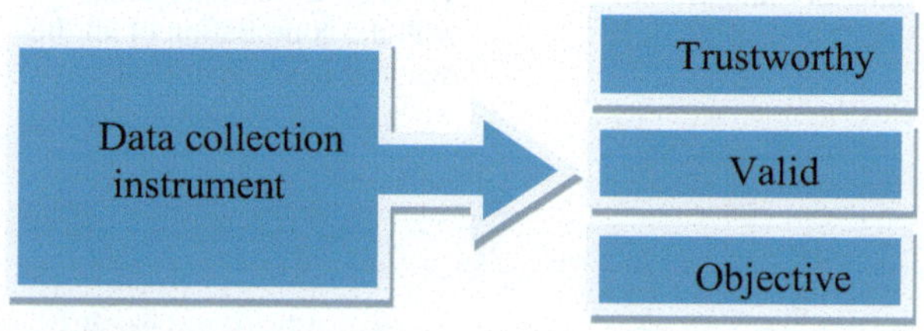

Fig. 10.4 Steps of the data collection methodology

In any case, the data collection method uses an instrument or a series of collection steps that can be used in a scientific investigation and this must meet three requirements: reliability, validity, and objectivity [28] as shown in Fig. 10.4.

Figure 10.2 shows us the trend of growth in public on the subject covered in this article, the highest peaks are observed between 2015 and 2017 with the greatest boom in 2016, because in these years the territorial development plan of the atlántic department or aligned with the national policy called Colombia Logistics Platform. Where the subject was studied largely by the different projects that began in the period described. For the period 2018 and 2019, stability is observed in line with the results publications and the specific incorporation of the terms Storage Loading and Intralogistics framed in Industry 4.0.

Therefore, the publications on the subject and observed in this publication are at the threshold from 2018 to 2019, covering a significant number of publications with reference to the terms intralogistic, Industry and Technology 4.0 and Storage Loading.

In this phase the information collection material is used the databases of the Autonomous University of Nuevo León—Monterrey and the Simón Bolívar University of Barranquilla the articles of consultation come from magazines that belong to the multi-disciplinary and important areas of engineering, including: Science Direct, Springer, IEEE Xplore, Scopus, and repositories of the aforementioned institutions. The documents that showed a similarity and great relevance with the topics involved in the present study with respect to industry and technologies 4.0, Storage Loading and Intralogistics, as well as the conceptualization of these terms, the analysis of environments, systems and paper were examined of the supply chain as an important aspect of logistics in the study of intralogistics in the years 2014–2019. However, articles or publications prior to 2014 were not taken in the analysis samples.

Sampling Planning

For this literature review, the 57 scientific publications, published between 2014 and 2019, were examined. Fig. 10.5 shows the graph of the selection of articles, and it is possible to see the considerable increase in research on the subject planted in this study, so it is independent and relevant to know the applications that can be derived from the combination of technologies and tools that are part of Industry 4.0 and define areas of opportunity and interest. After a search in the longitudinal distribution of publications [29], it is shown that, as of 2014, the number of published works has multiplied over the years. Therefore, in the coming years, we anticipate another "increase" as a result of the opportunities offered by emerging technologies applicable to intralogistic activities and processes.

In sampling strategies, reference should be made as defined by Mendieta [30] that the number of subjects is not central to qualitative research, the central thing is to describe what strategy, what type of sampling will be done, and how it will be done, always according to the question and the research design. If done horizontally or in stages. The important thing is to describe how the strategy is built.

The researches consulted come from 3 continents to 8 countries see Fig. 10.6, in which proposals are generated in application with theoretical tools including simulations from the look of the industry 4.0, intralogistics, and Storage Loading, it shows the classification of countries with research in this area. It is noted that

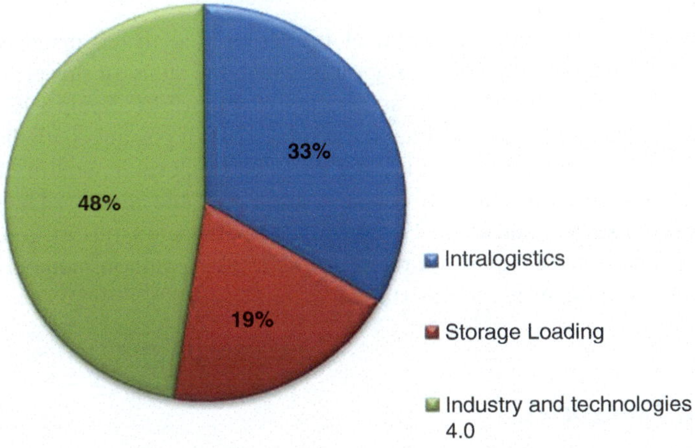

Fig. 10.5 Analysis of articles by categories

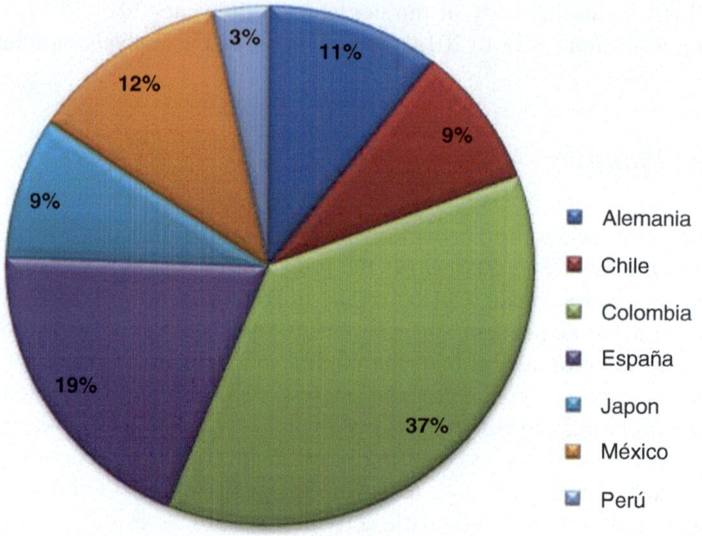

Fig. 10.6 Research consulted by country

most of the articles consulted are presented in research carried out by research is in Europe and Asia, establishing the beginning of this trend in these regions.

Figure 10.5 shows the relevance that thermal industry and technologies 4.0 48%, Intralogistics 33%, and Storage Loading 19% have among the publications consulted, so in logistical issues an interrelationship is denoted between the terms, but, taking into account in account for the component and characteristics of the logistics operators of the Atlantic department focus on Industry and technologies 4.0 in greater absorption.

Likewise, the correlation of search of the terms is analyzed determining that because of the characteristics of the region in which the study of this article arises, the articles of greater cohesion in the subject in Colombia 37%, Spain 19% the latter has a significant percentage according to the similarities it presents with the place of study, Mexico 12% and Germany 11% according to the development of the subject in its logistics operations through versatile and adaptable models. On the other hand there is Chile 9% and Peru 9% according to characteristics of geographical similarity. In the case of Japan 3% refers to the implementation methodology and therefore adapts to the needs of this study.

Choice of Categories

The categories for the articles were established in such a way where each of the articles is set in a single category, thus avoiding the interconnection between them.

Fig. 10.7 Industry Technology Group 4.0

The structuring of the sections is established during the research process and is carried out in accordance to the scenarios considered by the thematic axes of the literary review concerning industry and technologies 4.0, Storage Loading, and Intralogistics composed of sub-categories of grouping, see Fig. 10.7. These groups are diversified to establish a comprehensive system [31], considering intralogistics storage, loading and unloading of logistics operators towards their impact on organizational efficiency [32, 33].

The 57 articles reviewed were divided into blocks, which are the following: Intralogistics, Storage Loading, and Industry and technologies 4.0, the latter in turn divided into sub-blocks included in the topics of Simulation, Automation, and Sensorization of loads, collaborative Robotics, Big Data powered by M2M (Machine to Machine communication) and Predictive maintenance—predictive analysis. The Fig. 10.8 shows graph the articles of assignment to the year and percentage categories, in addition, you can see how there are some blocks that have been emphasized more in recent research and others that are under development.

Material Evaluation

The evaluation of scientific articles as a basis for analysis of this study is developed using a spreadsheet with a pre-established form, where the most important data of each article is organized by a number of features such as: publication name, authors, keywords, magazine name, volume number, date publication, pages and number of citations on the date of consultation, objective of the article, methodology, research approach, and conclusions.

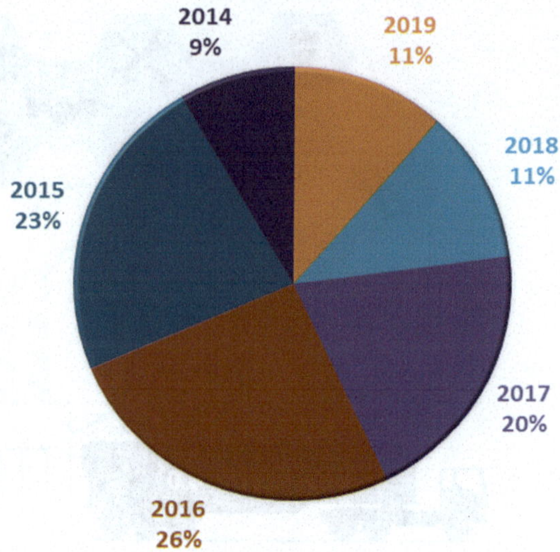

Fig. 10.8 Assignment of items per year

10.3 Discussion and Analysis of Information

To analyze the information resulting from the search in the databases and special sites in logistics and intralogistic dating, it allows a systematic review of the scientific articles determined in the methodology. However, this section is divided into three areas: Intralogistics, Storage Load, and Industry and Technology 4.0.

Intralogistics

Noachas [34] defines intralogistics as "the activities of the supply chain that are carried out within the company, in particular all those that have to do with the circulation and storage of goods within the company. The improvements or optimizations in these activities have a high impact on the business of the company."

Intralogistic activities are future-oriented, most of their supply chain control systems in material supplies, currently in the industry rely on centralized equipment management supplies [9] the emergence of new, effective and increasingly sophisticated information and communications technologies [10].

The prerequisite for intralogistics 4.0 is data transparency [11]. Intelligent optimization allows, for example, that all transport orders can be assigned efficiently to drivers and means of transport.

The Private Competitiveness Council (CPC) stated in accordance with the World Bank's Logistic Performance Index, during the 2018–2019 period. In 2018 Colombia obtained the highest qualification in its history in the Logistics Performance Index see Fig. 10.9. It advanced 36 positions compared to 2016 (he went from 94 to

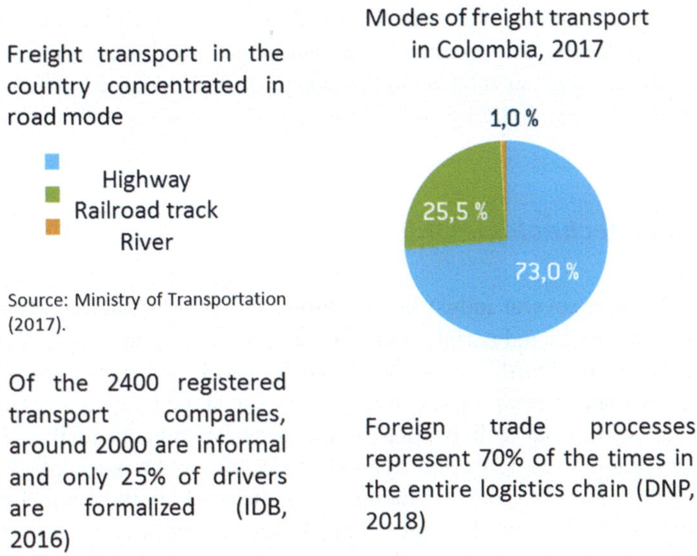

Fig. 10.9 Logistics Performance Index. Colombia and reference countries, 2018

58) and today he is fifth in Latin America [35]. However, there are still challenges in: efficiency and effectiveness of customs, quality of infrastructure, and competition and quality of logistics services [36].

Over time, the term has developed beyond its original sector and is now fully accepted in the field of logistics. Intralogistics is the last service link in the supply chain within the organization [37].

Intralogistic management systems today are compatible with the cloud, making its way to industry 4.0, facilitating the rapid and efficient response to the analyzed data, regardless of location [38]. The same applies to predictive maintenance, in which large amounts of data are used to obtain valuable information about plant maintenance [39].

Storage Loading

Storage Loading is a logistic term attributed to the Logistics of storage, storage load and in recent years has joined the terms used in the digital era and its passage towards the fourth revolution. López [40] defines the logistics of storage, as the activity that deals with the tasks of placing and saving the supplies received, keeping them in proper condition, as well as ensuring that the deposit of all these elements results in a positive way in the activity of the company.

Escudero [41] comments that "when we buy an item, most of the time we are not aware of its origin, who is its creator (manufacturer)," how it has reached the point of sale, how many stores or intermediaries has passed; these aspects define storage logistics.

Industry and Technologies 4.0

Throughout history, several industrial revolutions have been such as the meaning not only of changes in industrial, social, economic, and technological processes [42]. The road to the fourth era of the industrial revolution that is lived today is being called by many names, this is presented by the slope that has been emerging in accordance with what will be one of the greatest impacts in the change of deeper schemes and paradigms and dynamic [43], taking the name of: The era of Industry 4.0, the time of Connected Industry 4.0, the world in the Smart Industry or cyberindustry of the Future, the time of the Industrial Internet of Things, among others. Like the previous three revolutions, this new revolution is based on the application of new technologies to organizational processes both at the machinery and production level and along the value chain of the industrial process [44].

Since 2011 it is considered the beginning of a new industrial revolution [45]. This is structured under the trend of the characteristics of cyberphysical systems, industries or smart products, as well as interconnected cities. The new era consists of the combination of physical and tangible scenarios with information systems [24], to the degree of bringing operations to a level of synergy in real time. Technology based on networks and the Internet is considered [23] and the business world is visualized as an integral system under collaboration criteria. The approach considered in this revolution is intelligent and autonomous operations, however, the higher level must be evaluated by human reason for decision-making and sensitivity criteria [20]. This will involve the development of new applications and platforms that will facilitate in a simpler way the interconnection of the whole service and protagonists of the supply and transport chain [46].

However, the 4.0 technologies analyzed in this literary review were organized into four sub-categories: Simulation, Automation, and load sensing, Collaborative robotics, Big Data powered by M2M (Machine to Machine communication), and Predictive maintenance—predictive analysis and M2M communication (machine to machine).

Simulation

The simulation is the imitation of the operation of a real-world process or system over time [47]. Through its applications it allows us to observe the current production system and observe the relationship between networks with a cyberphysical environment and the coordination that it must have for the decision-making [48].

Simulation has been used for many years to develop CAD products and to assist in the engineering of CAM production. The novelty in the Industry 4.0 model lies in the use of real-time simulators of plant operators [49]. From the data captured at the exact time it happens, about the physical world is reflected in a virtual world and the following process is simulated to adjust it and optimize the parameters that condition it [50].

Automation and Load Sensing

As previously mentioned, automation considering the reality of Industry 4.0 will be analyzed in areas such as: Advanced factory level sensorization, Multistage and flexible manufacturing and Smart Manufacturing (cyberphysical systems (CPS)). The concept of advanced sensorization [51]. It refers to the technologies that, through the use of different sensors, allow to analyze and obtain data from different machines, operators or objects, non-invasively.

Collaborative Robotics

One way to visualize and strengthen industrial production is denoted in the processes that involve collaborative robotics, thus reducing errors in simple activities [52].

To achieve the efficiency of this chain, is necessarily determined relationships in the creation of collaborative robotics actions from the links of the characteristics of information technologies [53].

Big Data Powered by M2M (Machine to Machine Communication)

Data creation is established in all production systems [54], there are also proposals for autonomous organization and coordination, through an analysis in which the data is managed and filtered to generate useful information, the above sets the automation parameters of production systems, in quality and service [55].

Predictive Maintenance: Predictive Analysis and M2M Communication

Lughofer [56] defines predictive maintenance defines early integration of anomalies or fault detection, diagnosis and reasoning, prediction of remaining useful life (prognosis of failures), prediction of quality and self-reaction, as well as optimization, control and auto repair techniques (Table 10.2).

Table 10.2 Relationship between categories and sub-categories

Categories	Relationship
Intralogistics (A)	Carvajal [57]CD, Cibergenius [1]CD, DNP [2]BD, Muñoz [3]BC, Latina [4]BD, Logistic [6]CD, Román [8]BD, Schuhmaher [9]CD, Solar [10]CD, Schuh [11]BC, Marinko [13]CD, Neeraja [15]BC, Ltifi [16]CD, Madrigal [17]CD, Kaiser [18]CD, Ma [19]CD
Storage loading (B)	Dombrowski [20]AD, Hermann [21]AD, Qin [22]AC, Wolter [23]AD A B. [24]AC, Lacey [25]CD, Elo [26]CD, A Q.-P. [51]AC
Industry 4.0 (C)	Trends [31]CD, Toro [32]BD, A G. [33]BD, Eisenberg [34]AD construcción [35]CD, Investigación [36]AD, Mendieta [37]CD Schuh [38]AD, CPC [39]AB, CPC [40]AD, BID [41]BD U.R. [41]AB, Gómez [42]AD, López [43]AD Serrano [44]BD, Josefine [44]AD, Junco [45]BD
Technologies 4.0 (D)	Cortés [46]BD, K [47]AD, A B [48]BD, Wolter [49]AB Dombrowski [50]BD, Moreno [51]AC Ivanov [52]AB Guilera [53]AB, Garell [54]AC, Bagheri [55]CA, Schut [56]AC

10.4 Conclusion and Future Work

The intralogistics that begins to break into the scene of the organizations of Latin America, the Caribbean and especially the neo-Granada nation itself in its Caribbean region where its logistics operators are in the continuous search for tools that allow improving business efficiency and they have observed in the aspects of Intralogistics, Storage Loading, and Industry and Technologies 4.0, a roadmap for the fulfillment of this purpose.

Intralogistics from the literary point of view has much to apply and take advantage of 4.0 technologies to strengthen and improve Storage Loading in logistics operators, to make this leap to the so-called Intralogistics 4.0 a term that has been gaining ground in the studio literary and start-up within organizations as a tool for improving business efficiency. This article demonstrates a heterogeneous style of literature based on resources, contents, and concepts, shows a systematic review on advances regarding Industry 4.0, Intralogistics, and Storage Loading, including elements of 4.0 technologies. This literature review study uses 57 scientific papers that determine the following findings:

First of all, three categories of Intralogistics, Storage Loading, and Industry and Technologies 4.0 studies were identified, where the latter delimited five sub-blocks that are the backbone of incorporation of the 4.0 paradigm in this article. These sections are the central axis of the study and research, so you can see a panoramic view of their axes, as well as their importance of them.

Second, the selected articles apply to the respective study categories; however, an interconnection between the data of each block and sub-blocks is revealed, so that a systematic review of the characteristics raised is determined, in the same way the category with the greatest weight is identified, which reveals the requirements and steps necessary for successful implementation.

Third, it is revealed within the analysis that the category of sustained increase is the adoption and inclusion of technology related to Industry 4.0 within Storage Loading's intralogistic processes and activities, this is due to the fact that it represents the basis of industrial operations, this block is considered the nucleus, because it shows a slight dependence on the other blocks and sub-blocks established.

This literature review not only shows the three categories and applications of sub-categories, likewise, it establishes criteria concerning the management of intralogistic processes in terms of Industry 4.0 and its technology. In addition, the essential characteristics observed in the adaptation of the 4.0 environment allow to take a look at the future, the considerations of the logistics operators sector in terms of intralogistics and its application to the rest of the industry and the paradigm of technologies 4.0 becomes a less complicated paradigm if the starting point establishes the characteristic elementary criticisms for its implementation and adaptation. Determining the existence of an interrelationship between Storage Loading and intralogistics of logistics operators in the Atlantic department in which the application of technologies from Industry 4.0 improve systems and processes of Storage Loading allowing to impact organizational efficiency through the improvement of intralogistic processes, by establishing the storage loading as the backbone of intralogistics.

Thanks COMPSE 2019—third EAI International Conference on Computer Science and Engineering and Health Services, November 28–29, 2019. Mexico City, Mexico.

References

1. Cibergenius, *Cibergenius* (Honeywell International Inc., 2016) [En línea]. Available: https://cibergenius.com/codigos-barras-presente-toda-la-cadena/. Último acceso: 30 July 2019
2. DNP (Departamento Nacional de Planeación) (2015) [En línea]. Available: https://www.dnp.gov.co/Paginas/15-de-las-ventas-de-empresas-colombianas%2D%2Dse-va-en-log%C3%ADstica.aspx. Último acceso: 27 July 2019
3. S.G. Muñoz, *Departamento Nacioanl de Planeación* (2015) [En línea]. Available: https://www.dnp.gov.co/Paginas/15-de-las-ventas-de-empresas-colombianas%2D%2Dse-va-en-log%C3%ADstica.aspx. Último acceso: 27 July 2019
4. A. Latina, *Banco Mundial* (2017) [En línea]. Available: https://www.bancomundial.org/. Último acceso: 27 July 2019
5. D.N. da Silva, Optimización del Proceso de Innovación para Proyectos Internos en las Empresas. Scielo, **27**(3). (2016)
6. Logistic-editor, *Logistica Supply Chain Industria.* (2018) [En línea]. Available: https://revistadelogistica.com/tecnologia/latinoamerica-apuesta/. Último acceso: 27 July 2019
7. Arsam, *ARSAM logistic* (2019) [En línea]. Available: https://www.arsam.es/conoce-la-importancia-de-la-intralogistica-en-la-industria/. Último acceso: 24 July 2019
8. J. Román, *Industria 4.0: La Transformación Digital de la Industria* (Coddiinforme, 2016)
9. J. Schuhmaher, Development of a descriptive model for intralogistics as a foundation for an autonomous control method for intralogistics systems. Procedia Manuf. **23**, 225–230 (2018)

10. L.E. Solar, *Mejorando la Gestión de la Logística de Almacenamiento Inbound en cd* (Media Parner Logistec, 2015) [En línea]. Available: https://www.revistalogistec.com/index.php/equipamiento-y-tecnologia/almacenaje/item/2389-mejorando-la-gestion-de-la-logistica-de-almacenamiento-inbound-en-cd. Último acceso 13 July 2019
11. L. Schuh, *MMLOGISTIK* (2017) [En línea]. Available: https://www.mm-logistik.vogel.de/mit-digitalisierung-und-automation-zur-intralogistik-40-a-593552/. Último acceso: 27 June 2019
12. L. Francisco Marcelo, *Análisis y propuestas de mejora de sistema de gestión de almacenes de un operador logístico* (Pontificia Universidad Católica del Perú, 2014)
13. M. Marinko Maslarić, Logistics response to the industry 4.0: The physical internet. De Gruyter Open **6**, 511–517 (2016)
14. S.I. Chiavenato, *Planeación estratégica* (McGraw-Hill, Sao Paulo, 2017), pp. 4–14
15. B. Neeraja, M. Mehta, A. Chandani, Supply chain and logistics for the present day business. Proced. Econ. Financ. **11**, 665–675 (2014)
16. M. Ltifia, J. Gharbib, The effect of logistics performance in retail store on the happiness and satisfaction of consumers. Proced. Econ. Financ. **23**, 1347–1353 (2015)
17. S.A.M. Moreno, N.D.M. Ceballos, Vehículos de guiado autónomo (AGV) en aplicaciones industriales: una revisión. Rev. Politecn. **15**(28), 117–137 (2019)
18. J. Kaiser, Diseño de flujo de valor VSD enfoque morfologico en la cadena de valor. Springer **3**(15), 64–79 (2017)
19. Y. Ma, Storage systems management based on big data - Take three common storage systems as examples, de *15th International Conference on Service Systems and Service Management (ICSSSM)* (Hangzhou, China, 2018)
20. U. Dombrowski, T. Wagner, Mental strain as field of action in the 4th industrial revolution. Proc. CIRP **17**, 100–105 (2014)
21. M. Hermann, T. Pentek, B. Otto, Design principles for Industrie 4.0 scenarios, de *49th Hawaii international conference on system sciences (IEEE)* (Hawaii, 2016)
22. J. Qina, Y. Liua, R. Grosvenora, A categorical framework of manufacturing for industry 4.0 and beyond. Proc. CIRP **52**, 173–178 (2016)
23. M. Wolter, Industrie 4.0 und die Folgen für Arbeitsmarkt und Wirtschaft, SzenarioRechnungen im Rahmen der BIBB-IAB-Qualifikations-und Berufsfeldprojektionen. IAB Forschungsbericht **8**, 1–69 (2015)
24. A. Botthof, E.A. Hartmann, Zukunft der arbeit im kontext von autonomik und industrie 4.0, in *Zukunft der Arbeit in Industrie 4.0*, ed. by A. Botthof, E. A. Hartmann, (Springer, Berlin, 2015), pp. 3–8
25. A. Lacey, D. Luff, *Qualitative Data Analysis*, vol 46 (Trent focus Sheffield, Nottingahm, 2001)
26. S. Elo, M. Kääriäinen, O. Kanste, T. Pölkki, K. Utriainen, H. Kyngäs, Qualitative content analysis: A focus on trustworthiness. SAGE Open **1**(4), 1–10 (2014)
27. A. Quintana, *Metodología de Investigación Científica Cualitativa* (Tópicos de Actualidad, Psicologia, 2006), pp. 47–84
28. l. construcción, *Pautas generales para la realizar seminarios de investigción en ciencias de los datos* (Universidad de Chile, Santiago, 2016)
29. Dirección de Fomento a la Investigación, *Política para mejorar la calidad de las publicaciones científicas nacionales* (Colciencias, Bogotá, 2016)
30. G. Mendieta Izquierdo, Informantes y muestreo en investigación cualitativa. Investig. Andina **17**(30), 1148–1150 (2015)
31. Trends Magazine, *Industry 4.0 and the US manufacturing renaissance., Trends e-mag* (Audio-Tech Business book summaries, 2015), pp. 4–10
32. C. Toroa, I. Barandiarana, J. Posadaa, A perspective on knowledge based and intelligent systems implementation in industrie 4.0. Comput. Sci. **60**, 362–370 (2015)
33. A. Gilchrist, *Middleware Industrial Internet of Things Platforms. Industry 4.0: The Industrial Internet of the Things* (Springer, Berlin, 2016), pp. 153–160
34. G.N. Eisenberg, *Intralogística el corazón de la logística empresarial* (Editorial Académica Española, 2017)

35. CPC (Conseo privado de competitividad), *Mejorando la Competitividad en Colombia* (2018) [En línea]. Último acceso: 30 June 2019
36. CPC (Consejo privado de competitividad), *La competitividad en Colombia* (2017) [En línea]. Último acceso: 28 June 2019
37. BID (BAnco Interamericano de Desarrollo) (2017) [En línea]. Available: https://www.iadb.org/es/paises/colombia/perspectiva-general. Último acceso: 17 May 2019
38. U. R. a, Manutención y almacenaje: logística, distribución, transporte. **532**, 34–35 (2018)
39. S. Gómez, *Técnicas de Mantenimiento Predictivo. Metodologia de Aplicación en las Organizaciones* (Universidad Católica de Colombia, Bogotá, 2017)
40. F. López, *Economipedia, Logistica de almacenamiento* (2019) [En línea]. Available: https://economipedia.com/definiciones/logistica-de-almacenamiento.html. Último acceso: 28 July 2019
41. E. Serrano, *Logistica de Almacenamiento* (Ediciones Paraninfo, España, 2014)
42. C. Josefina, *Historia de la Industria 4.0* (Logicbus, 2019)
43. M. Junco, *El mundo ya está viviendo la Cuarta Revolución Industrial* (P0rtafolio.co, 2017)
44. Y. Cortés, El entorno de la industria 4.0: implicaciones y perspectivas futuras. Concienc. Tecnol. **54**, 33–45 (2017)
45. K. Simonis, Y.-S. Gloy, T. Gries, INDUSTRIE 4.0-automation in weft knitting technology. Mater. Sci. Eng. **1**, 1–10 (2016)
46. A. Moreno, G. Velez, A. Ardanza, I. Barandiaran, Á.R. de Infante, R. Chopitea, Virtualisation process of a sheet metal punching machine within the industry 4.0 vision. Int. J. Interact. Des. Manuf. (Springer) **2**(11), 365–373 (2017)
47. M. Scanlan, *Simulación de Manufactura para Industria 4.0.* (USA Engineering, 2017) [En línea]. Available: https://www.engusa.com/es/posts/simulacion-de-manufactura-para-industria-4-0. Último acceso: 31 July 2019
48. M.I. Ivanov, Schedule coordination in cyberphysical supply networks industry 4.0. IFAC-PapersOnLine **12**(49), 839–844 (2016)
49. L. Guilera, *Logisnet Cadena de Suministros* (Logisnet, 2019) [En línea]. Available: http://logisnet.com/. Último acceso: 29 July 2019
50. A. Garell, *La Industria 4.0 en la sociedad digital* (Marge Book, Barcelona, 2019)
51. A. a, *Estado del Arte de Automatización y Robótica* (Igape, Galicia, 2017)
52. L. Javier Darío Fernández, La industria 4.0: Una revisión de la literatura. Actas de Ingeniería **3**, 222–227 (2017)
53. L. Bagheri, Cyber-physical systems architecture for self-aware machines in industry 4.0 environment. IFAC PapersOnLine **3**(48), 1622–1627 (2015)
54. J.H. Schut, Keeping up with smarter machines. Plast. Eng. **52**, 10–16 (2016)
55. Z. Wang, Towards smart factory for industry 4.0: A self-organized multi-agent system with big data based feedback and coordination. Comput. Netw. **101**, 158–168 (2016)
56. M. Lughofer, *Predictive Maintenance in Dynamic Systems* (Springer, Cham, 2019)
57. Carvajal, *Logística Supply Chain - Industria* (2016) [En línea]. Available: https://revistadelogistica.com/logistica/intralogistica-la-herramienta-clave-de-la-gestion-operacional/

Chapter 11
Secure Key Distribution Prototype Based on Kerberos

Abigail Verónica Chantes Barrios and Francisco de Asís López Fuentes

Contents

11.1 Introduction 143
11.2 Related Work 144
11.3 Architecture 145
11.4 Implementation 149
11.5 Evaluation and Analysis 152
11.6 Conclusions 153
References 154

11.1 Introduction

Over the years, people have had a greater dependence on technologies to maintain their lifestyle. This dependence has been more marked by advances in TICs, which have spread the demand for distributed applications. Using distributed applications, users can share resources or distribute tasks more easily. However, these tasks must be performed safely in order to protect the integrity and confidentiality of the data used by these distributed applications. Due to communications being opened in distributed systems, different security risks such as viruses, Trojans, or Ransomware are introduced [1]. Therefore, security in TICs is an important point to consider. This security must cover logical and physical procedures to prevent, detect, and correct processes of misuse, as well as to resist different types of attacks. Researchers in both academia and industry are making efforts to innovate in secure distribution architectures to meet the challenges of a rapidly growing market.

In this paper, we present a prototype to distribute keys securely as an extension of the Kerberos authentication protocol. Seeing that Kerberos does not provide authentication between the servers that integrate it [2], since it is an authentication system

A. V. C. Barrios · F. de Asís López Fuentes (✉)
Department of Information Technology, Universidad Autónoma Metropolitana – Cuajimalpa, Mexico City, Mexico
e-mail: 2143066533@alumnos.cua.uam.mx; flopez@correo.cua.uam.mx

P. Vasant et al. (eds.), *Data Analysis and Optimization for Engineering and Computing Problems*, EAI/Springer Innovations in Communication and Computing,
https://doi.org/10.1007/978-3-030-48149-0_11

designed for open network computing environments. This document proposes a prototype for the authentication of the multiple servers that integrate Kerberos. This solution can provide reliability to a server system, as it has better authentication between servers and reduces the risk of possible attacks during key distribution. During the authentication process, the implementation of our prototype combines the AES (Advanced Encryption Standard) and RSA (Rivest, Shamir, and Adleman) algorithms, in order to provide better security, integrity, and authentication between them.

The rest of this paper has been organized as follows. In Sect. 11.2 we give a brief description about related works. Mainly some Kerberos' vulnerable points are explored. Section 11.3 shows the new scenario that was created to give a solution to the problem that had been detected. In this case, it is about the authentication of Kerberos servers. Each of the stages in the new proposal for the Kerberos protocol is explained. Implementation of our prototype is described in Sect. 11.4. This implementation intends to provide greater security when wanting to authenticate servers using distributed session keys (KDC) in order to reduce the risk of possible attacks. Evaluation of a prototype and the obtained results are described in Sect. 11.5. Our conclusions and some ideas as future work are presented in Sect. 11.6.

11.2 Related Work

Kerberos is an authentication protocol widely used in a network based on a third trusted [3]. Kerberos was developed as part of Project Athena at the Massachusetts Institute of Technology [4]. Kerberos version 5 is the latest version and it was published in 2005 [5]. The original Kerberos model is based in part on Needham and Schroeder protocol, and on several modifications proposed by Denning and Sacco [14]. However, the Kerberos v5 protocol was published due to new requirements and features that are not available in version 4. This implementation was directed by John Kohl and Theodore. In spite of several improvements introduced in Kerberos several improvements are still in place. For example, Hu et al. [6] propose a key exchange scheme based on Kerberos Diffie-Hellman-DSA. In this work, authors use the original Diffie-Hellman protocol for the distribution of public keys, which allows two users to exchange keys with security. However, this protocol does not provide an authentication mechanism for exchanged public keys. The main objective in this work is to improve the performance and efficiency of Kerberos, making a mutual authentication between the servers with the exchange of keys knowing the possible attacks to which the Kerberos protocol is vulnerable. Another work related to Kerberos authentication has been proposed by Pérez-Méndez et al. [7]. In this case, authors emphasize that, before an end user accesses any application service located in the service provider, it must be authenticated by the identity provider where the end user is registered. In this part, the authentication and authorization information and related decisions are sent to the service provider, responsible for validating them through a set of trust established between the identity provider and the service provider. This is for the purpose of providing a better security among the requesting servers.

11.3 Architecture

The basic Kerberos protocol is functional for the following reasons [8]. However, Kerberos does not perform authentication between its servers (AS, TGS, SS). Therefore, a scheme to distribute keys between the different Kerberos servers is required. Our proposal considers an arbitrator or KDC (Key Distribution Center) as a validator of authentication. Figure 11.1 shows this architecture and a preliminary description of how this protocol operates [9]. A detailed description of our model is illustrated in Fig. 11.2, using a UML diagram. Figure shows the required calls and procedures.

The initial version of the proposed protocol (version 1) consists of seven components: the client (C), the authentication server (AS), the Ticket Granting Server (TGS) and the service server (SS), the symmetric encryption mechanism Advanced Encryption Standard (AES), asymmetric cryptography system RSA (Rivest, Shamir, and Adleman) and an arbitrator in charge of enabling validation between the Authentication Server (AS) and the Ticket Granting Server (TGS) or between the Ticket Granting Server (TGS) and the service server (SS).

Step 1: Any of the SA, TGS, or SS servers want to make a request to authenticate with any other. Therefore, each server must send its ID and password, so that the referee can assign a session key to each server.

Step 2: When referee receives a request from any server, this generates a key using the AES encryption algorithm.

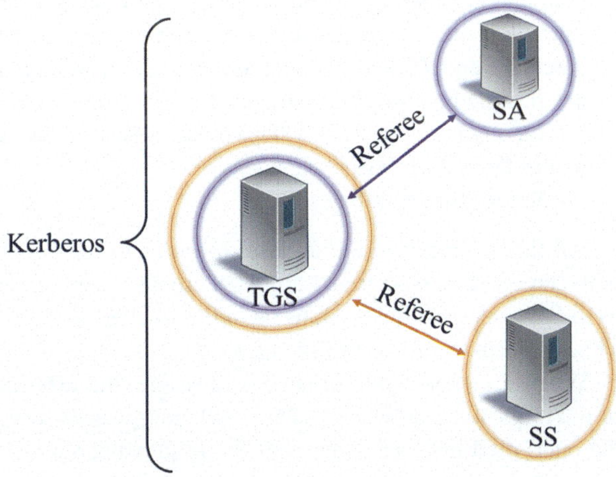

Fig. 11.1 Description of the possible scenario

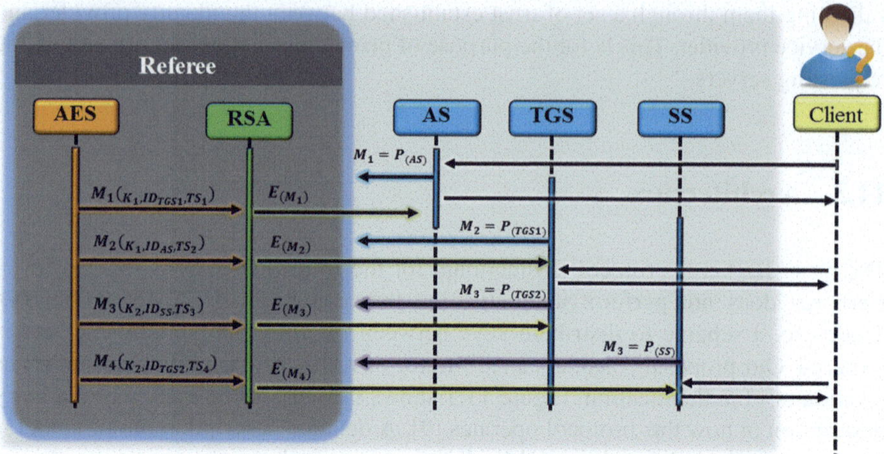

Fig. 11.2 Architectural Model (version 1)

Step 3: AES returns the message (m_n), which contain: the generated key (K), the ID to identify which server sent it, and the timestamp (TS). Creation of timestamps is based on the distance to which the servers are located. That messages have an assigned time to do the communication between servers.

Step 4: The message to be exchanged is encrypted using the RSA algorithm, which generate the keys, and encrypt and decrypt the message.

Step 5: Once the message arrives and it is encrypted correctly, the referee sends to the server this message ($E_{n(K_n, ID_X, TS_n)}$).

Where:

- E: Encrypted message n
- K: Assigned key of the server, for example: for *AS* and *TGS*, they will have assigned the same K, because these servers need to authenticate between them.
- ID_x: Identifier of the server to be communicated. For example: communication enters *AS* and *TGS*, for *AS* the assigned *ID* would be that of TGS_1, while for TGS_1 the assigned *ID* would be that of *AS*. The same would happen between TGS_2 and *SS*, for TGS_2 the assigned *ID* would be that of *SS*, while for *SS* the assigned *ID* would be that of TGS_2.
- *TS*: Server timestamp (n).

All this is stored in a database, in order to have a record of the communication between servers. As a first step, AES creates a session key for after this be assigned to the two servers that wish to be authenticated. Also their ID and ST are included. This scenario is shown in Fig. 11.3.

When a server wants to send a message, the arbitrator looks for the public key of the requesting server (*AS*, *TGS*, *SS*), encrypts its message with this key and, once the encrypted message arrives at the requesting server, decrypts it using its private

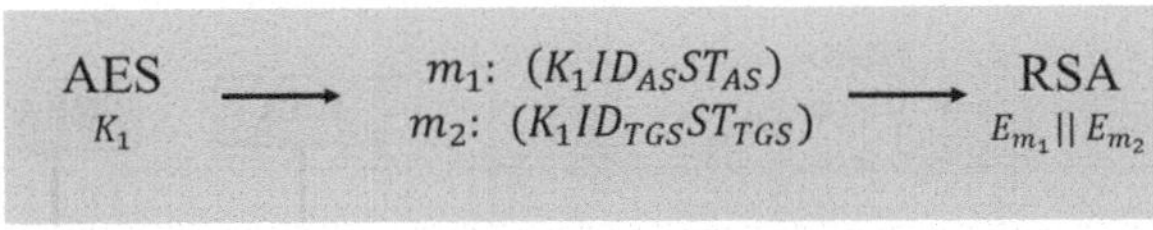

Fig. 11.3 Description of referee

Table 11.1 Advantages and disadvantages of encryption and decryption of keys

$K_R \rightarrow K_U$	$K_U \rightarrow K_R$	Symmetric key
Only the referee can encrypt YES authentication	The server cannot open the message with the private key of the referee NO authentication	Problem to distribute keys NO authentication
Everyone knows the public key NO confidentiality	Everyone knows the public key NO confidentiality	Problem to distribute keys NO confidentiality

key. Table 11.1 shows the cases that were found for this first version of the protocol are as follows:

- Case 1: Using Public Key ($K_R \rightarrow K_U$)
  If message was encrypted with the private key (K_R) and decrypted with the public key (K_U), then only the referee can encrypt the message, therefore, it provides authentication. However, this also implies that everyone knows the referee's public key and can obtain the session key, but this does not provide confidentiality.
- Case 2: Using Private Key ($K_U \rightarrow K_R$)
  If message was encrypted with the public key (K_U) and decrypted with the private key (K_R), then the receiver cannot open the message because the referee cannot disclose its private key, therefore it does not provide authentication. However, this also implies that everyone knows the public key and could decrypt the message, therefore it does not provide confidentiality.
- Case 3: Using Symmetric Key
  If message was encrypted with the session key using a symmetric key both on the referee side as the requesting server side, then we have the problem how to distribute the key.

Therefore, another version of the UML diagram of the already improved prototype (version 2) was created for the authentication of the servers that are developed in this protocol. This new approach is shown in Fig. 11.4.

However, this new version is based on the RSA protocol, so that each server (SA, TGS, SS) is able to generate their session keys (K_U, K_R), with them it will be possible to encrypt and decrypt the message, just like the referee. But, they do not exchange their keys at any time. For example, Fig. 11.5 shows how the public and private keys of the referee and AS are generated. The referee would be assigned a Public Key 1 (K_{U1}) and a Private Key 1 (K_{R1}), whiles the AS would be assigned a Public Key 2 (K_{U2}) and a Private Key 2 (K_{R2}).

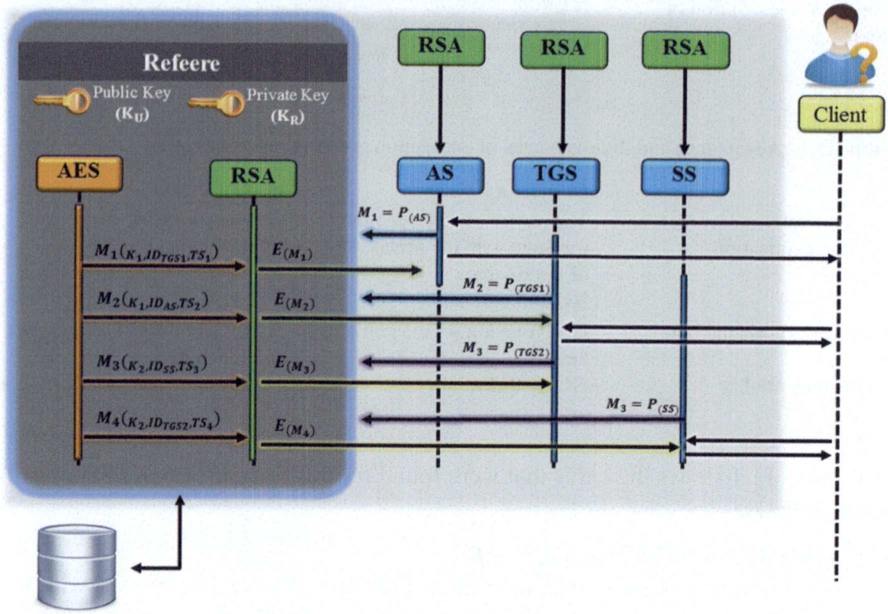

Fig. 11.4 Architectural model (version 2)

Fig. 11.5 Assignment of keys for the referee and AS

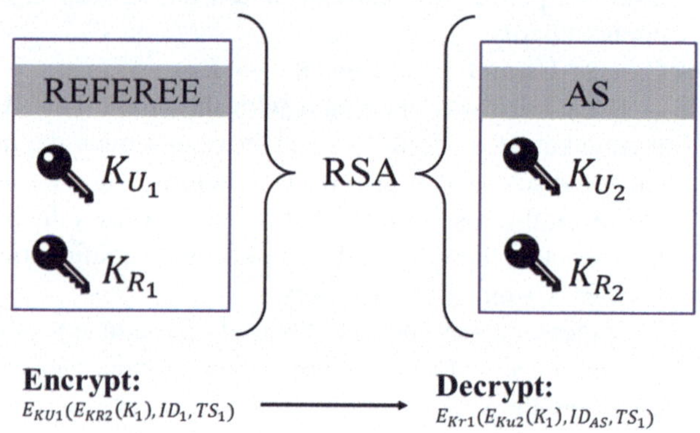

For confidentiality, a message to AS can be encrypted with the public key of AS in the referee side, and this message can be decrypted with the private key of AS in the AS side. While for the authentication a message is signed by AS using its private key and referee can use the public key to open this message. The purpose of this scheme is to provide confidentiality and authentication.

11.4 Implementation

A prototype using our proposed key distribution scheme based on Kerberos has been implemented in our lab. Our implementation is carried out in the Operating System Ubuntu 16.04, and the implementations of our algorithms are programmed in language C. In the next sections we describe the different steps to build our prototype.

AES

In cryptography, AES (Advanced Encryption Standard) is a symmetric block cipher system chosen by the government of the EE.UU. AES was developed by Joan Daemen and Vincent Rijmen [10]. AES is one of the most popular algorithms used in symmetric cryptography. The AES encryption is by blocks, which means that data is divided into a segment of 16 bytes (128 bits), and each segment can be seen as a 4 × 4 block or matrix, which is called a state.

In the improved UML diagram, only the AES algorithm is called once, it is at the moment of generating a key, which will be shared by the compatible servers (see Fig. 11.6).

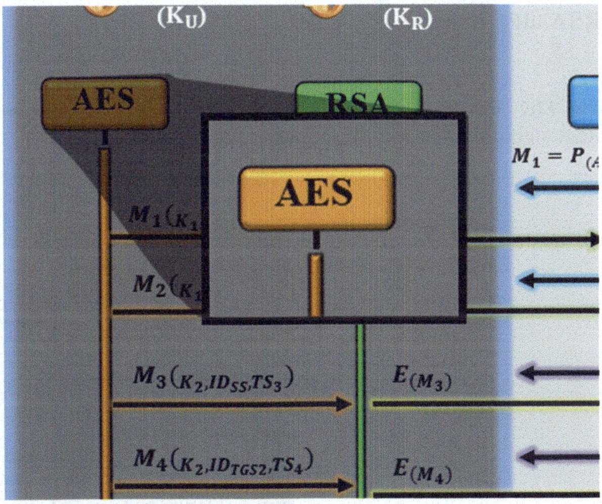

Fig. 11.6 Call of the AES algorithm

RSA

In cryptography, RSA (Rivest, Shamir, Adleman) is a cryptographic key system published in 1979 [11]. The RSA algorithm consists of three steps: generation of keys, encrypt message, and decrypt message, which are described below [12]. In the improved UML diagram of the prototype (version 2), it calls the RSA algorithm on two occasions.

- *First call:* It has been sent to be able to encrypt the information received to the referee by the servers (SA, TGS, SS) and to be able to authenticate itself in a secure way. The data that the servers send would be their key ID and the server with which it wants to be authenticated (see Fig. 11.7.). In this call the algorithm encrypts the message which consists of information extracted from a database. Once the database authenticates the request it will extract the ID of the server. The message to be encrypted will be the key generated by the AES plus the server ID to authenticate plus the timestamp ($KeyAES + IDServerA + Timestamp$).

- *Second call:* It is done to have a better security at the moment of encryption or decryption of the message with the assigned keys. This is due to disadvantages found, which are shown in Table 11.1 (see Fig. 11.8).

For this encryption algorithm two files were created "rsa.c" and "functions.h." In the .c file it sends to call the functions to generate keys, encryption and decryption. In the file .h is composed of functions, which are required for the operation of the RSA algorithm, where:

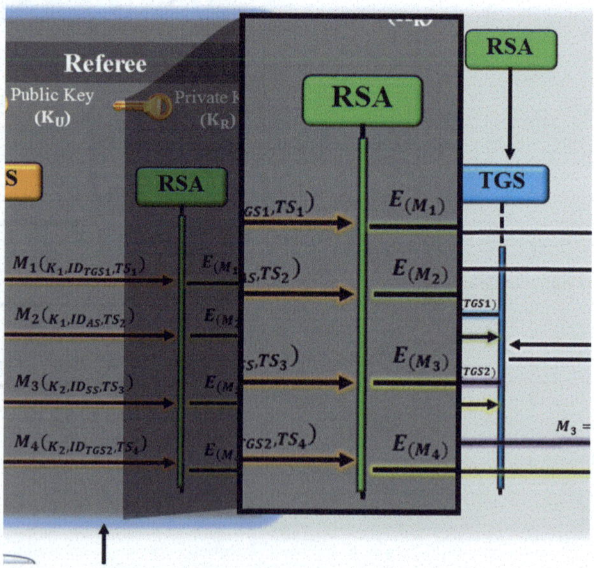

Fig. 11.7 First call of the RSA algorithm

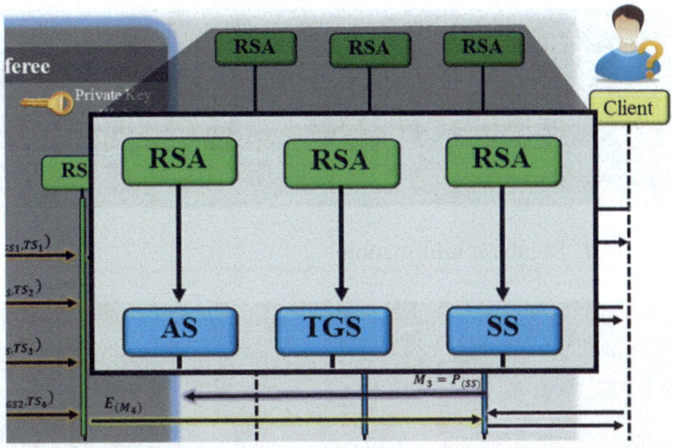

Fig. 11.8 Second call of the RSA algorithm

- *int generate():* This function generates two random primes (p and q). As long as the values of p and q are greater than 2 and less than 1000, it generates two prime random numbers.
- *void encrypt():* This function is used to encrypt the input message. Opening a file with the name "encrypted.txt" to write the encrypted message.
- *void decrypt():* This function is used to decrypt the message. By calling the function AEM (Modular Exposure Algorithm), decrypting each of the encrypted characters.
- *int maxCD()*: This function obtains the Maximum Common Divisor between two numbers.
- *int mod_ reverse():* In this function, the inverse of the variables e and *phi* are calculated.
- *int prime():* This function checks whether the entry number is a prime number or not.
- *int valueE():* This function calculates the value of "e" in a random way.
- *int AEM():* In this function the Modular Exhibition Algorithm is implemented.
- *void delete_entry ():* This function clears the text array.

MySQL

MySQL, the most popular Open Source SQL database management system, is developed, distributed, and supported by Oracle Corporation [13]. In the improved UML diagram of the prototype (version 2), the arbitrator consults this database created each time the servers request to know if the communication is secure. This database consists of six columns (id_Servers, server_name,IP, puert, time_Stamp, AES, and id_connection) and four rows (SS, AS, GTS1, and GTS2) to store the information of the servers that make up Kerberos (see Fig. 11.9).

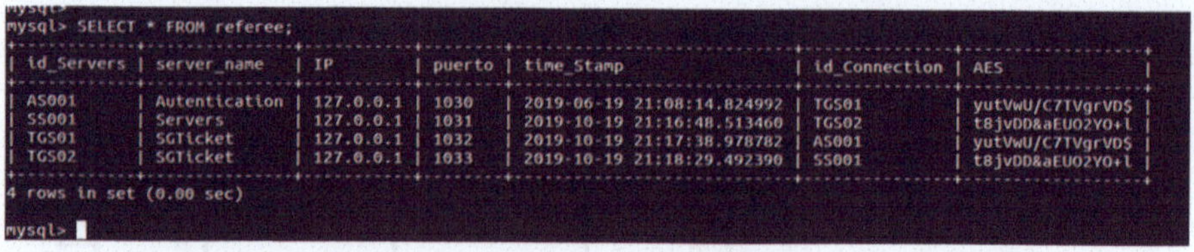

Fig. 11.9 Database information

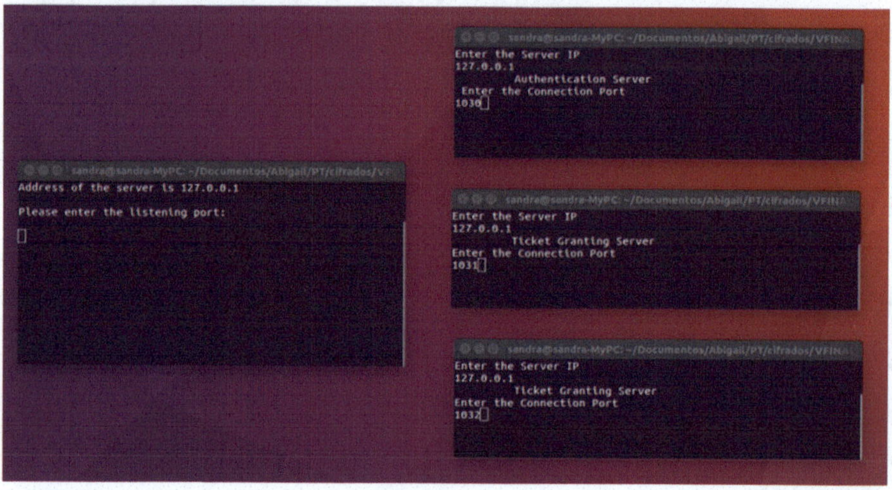

Fig. 11.10 Ports assigned to each server

Servers

To implement the test servers, programming language C was used. All servers are connected in the IP (127.0.0.1), with different ports (see Figs. 11.10).

11.5 Evaluation and Analysis

For the execution of the programs knowledge of the Linux operating system is required. For the following evaluations of the algorithms for the implementation an alternative menu is shown. This is only for the purpose of observing the executions of the RSA algorithm. When establishing a communication between the servers, they must know the IP address and the port of the server with which they wish to communicate (see Fig. 11.10). Once the communication is established, the server calls to the MySQL database (referee). The referee's table will be used to authenticate the server's first request to the referee. For this, the server sent its ID and password so that the referee rebids if that server is authentic (see Fig. 11.11).

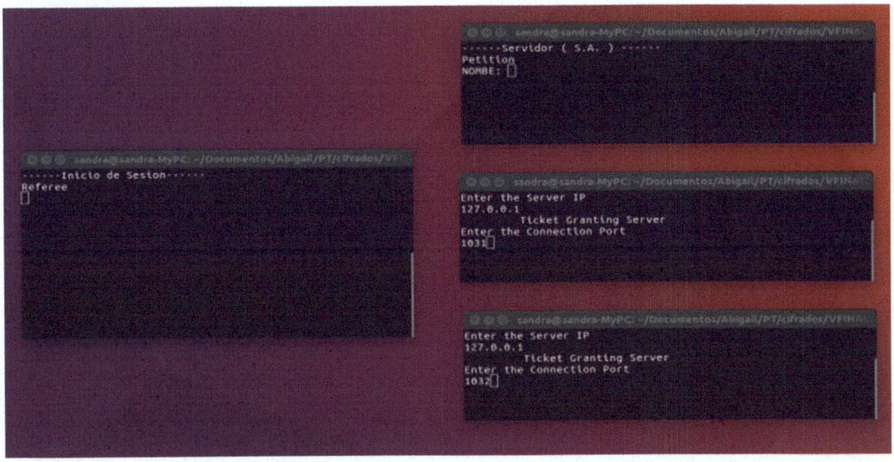

Fig. 11.11 Petition between AS and referee

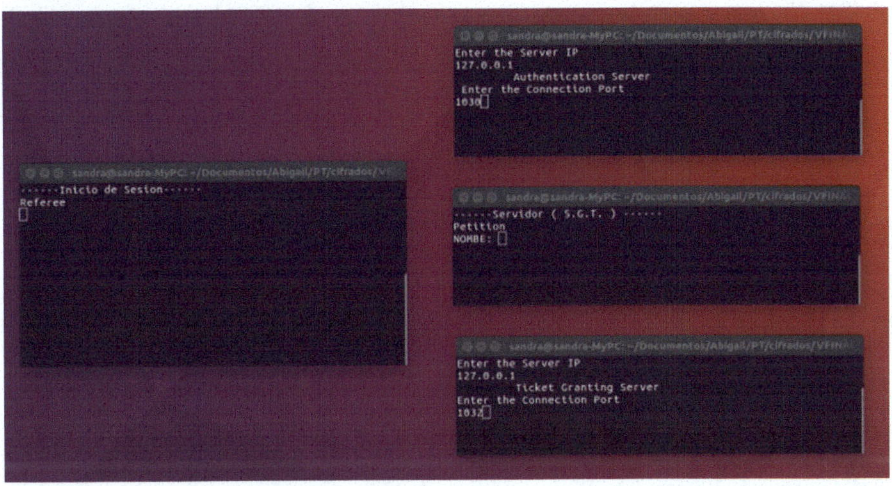

Fig. 11.12 Petition between TGS and referee

If the information is authentic, then follow the process of communicating Kerberos servers (see Fig. 11.12). If it is not correct, then the request is denied.

11.6 Conclusions

Nowadays, technologies are evolving more and more rapidly and society suffers the risk of attacks, as in social networks, with the main objective of obtaining information from users who use them (data as: emails, addresses of house, telephone numbers, etc.). On the other hand, it is also common that when users request a web service they do not have the security that this address is authentic Therefore, we must have several applications to ensure that a server is secure and that software resources are used only for the purposes that were created. When users make their requests and they obtain access code, they feel more security to surf the web and to entrust their data without any fear. Our prototype presented in this paper has a main objective to provide greater security within the servers that communicate in order to avoid possible attacks. This goal is reached by our prototype by involving encryption techniques with passwords, and timestamps during key distribution between servers.

References

1. Symantec Homepage, https://www.symantec.com/es/mx/security-center/threat-report. Last accessed 30 June 2018
2. IBM Homepage, http://www.vsantivirus.com/vul-kerberos5-060904.htm. Last accessed 30 June 2018
3. Cisco Homepage, https://www.cisco.com/c/dam/global/es_mx/solutions/pdf/reporte-anual-cisco-2018-espan.pdf. Last accessed 7 Nov 2018
4. Cisco Homepage, https://www.cisco.com/c/es_mx/support/docs/securityvpn/kerberos/16087-1.html. Last accessed 11 July 2018
5. C. Neuman, T. Yu, S. Hartman, K. Raeburn, *The Kerberos Network Authentication Service (V5)*, RFC 4120 (2005)
6. Z. Hu, Y. Zhu, L. Ma, An improved Kerberos protocol based on Diffie-Hellman-DSA key exchange, in *18th IEEE International Conference on Networks*, Singapore, Singapore (2012), pp. 400–404
7. A. Pérez-Méndez, F. Pereñíguez-García, R. Marín-López, G. López-Millán, Out-of-band federated authentication for Kerberos based on PANA. Comput. Commun. **36**, 1527–1538 (2013)
8. MIT Kerberos Consortium Homepage, https://www.kerberos.org/software/whykerberos.pdf. Last accessed 23 Sep 2018
9. Cisco Homepage, http://www.vsantivirus.com/vul-kerberos5-060904.htm. Last accessed 30 June 2019
10. J. Daemen, V. Rijmen, *The Design of Rijndael: AES - The Advanced Encryption Standard*. (Springer, 2002). ISBN 3-540-42580-2
11. University of Chicago Homepage, http://www.math.uchicago.edu/~may/VIGRE/VIGRE2007/REUPapers/FINALAPP/Calderbank.pdf. Last accessed 10 July 2018
12. J. Johnson, B. Kaliski, *Public-Key Cryptography Standards (PKCS) #1: RSA Cryptography Specifications Version 2.1*. www.ietf.org. (Network Working Group, 2016)
13. MySQL Homepage, https://dev.mysql.com/doc/refman/8.0/en/what-is-mysql.html. Last accessed 29 May 2019
14. D. Denning, G. Sacco, Timestamps in key distributed protocols. Commun. ACM, **24**(8), 533–535 (1981)

Chapter 12
A Decision Making Approach Using Fuzzy Logic and ANFIS: A Retail Study Case

Tomas E. Salais-Fierro, Jania Astrid Saucedo Martínez, and Blanca I. Pérez-Pérez

Contents

12.1 Introduction 155
12.2 Literature Review 156
12.3 Methodology 161
12.4 Inventory Policies Development 167
12.5 Conclusion 170
References 171

12.1 Introduction

In the current economy, an adequate supply chain management is mandatory to guarantee the survival of the industries, allowing them to increase their competitiveness.

The management and control of the companies resources is vitally important to achieve competitiveness in these companies. This implies that organizations must have a control inventory which allows to fulfill with the customer demand, keeping the control stock handling costs at an affordable level for the company [1].

The potential improvement of the inventory management collaborates to the income increasing, costs reduction, and a significant increment in the customer satisfaction. Hence, the utilization of inventory policies is significantly useful and is a study motivation for the logistic specialists [2]. In this way, the development of the management policies has kept the attention of the engineering researchers and the logistics chains. The main objective of an adequate inventory policy is to acquire a desired inventory level, avoiding complex purchasing orders and the stock variability [3].

T. E. Salais-Fierro (✉) · J. A. Saucedo Martínez · B. I. Pérez-Pérez
Universidad Autónoma de Nuevo León, San Nicolás de los Garza, Mexico
e-mail: tomas.salaisfr@uanl.edu.mx; jania.saucedomrt@uanl.edu.mx; blanca.perezprz@uanl.edu.mx

P. Vasant et al. (eds.), *Data Analysis and Optimization for Engineering and Computing Problems*, EAI/Springer Innovations in Communication and Computing,
https://doi.org/10.1007/978-3-030-48149-0_12

This paper is divided into five sections. The following section refers to the literature revision prepared to describe some cases where interesting situations are exposed of the enterprises related to the supply chain and how have been resolved, approaching points like the inventory classification ABC, the forecasting analysis, and the fuzzy logic. Then in the next section, the study case is shown and the main problem is described. In the third part of the paper, the selected methodology is presented in which the solution of the case is based and the ANFIS tool is described, the one used to solve the main issue. In the fourth section, its describes the method in which the experiment was performed describing the obtained results based on the study case, going by some of the most representative products belonging to the inventory type A classification and with a high rotation index. Finally, the conclusions of the current project are presented and the recommendations for future research are highlighted.

12.2 Literature Review

In literature, a great interest is seen in this area, because it is a significant part of the performance of a company. As an example of the mentioned above, in this section a short review of different works is executed, related with this topic and the utility of the proposed solution tools.

A supplying policy consists on the decisions respecting on when and how much to reorder. These policies may adopt a variety of forms. Then, two types are mentioned by Chopra and Meindl [4]:

Continuous review: The stock is revised continuously. The time between the orders may fluctuate, considering the variable demand.

Periodic review: The status of inventory is inspected to regular ranges, then an order is requested to increase the stock level until to a specified one.

ABC Method

When it is required to manage a wide number of products, it is suggested the aggrupation and the elaboration of stock policies corresponding to each one of the resulting groups [5]. The ABC method is useful to get a better follow up of the stock, taking reference on the sales and the products volume [6]. The classification with the ABC inventory methodology is often applied in the industry [7, 8], which was created by GE [8] in 1950. Based on ideas of Pareto [9], it is concluded approximately the 20% of the stock contributes to the company the 80% of the total sales and incomes of the organization; or well-known as the 80–20 rule. The stock is classified into three main groups which are A, B, and C. The 20% of the items represents the A group, the next 30% to the group B, and the lasting 50% to the group C [7]. This allows us to focus our attention, firstly on the products that contribute to higher earnings (products type A) and to serve to the rest of the classification based on the company chances, taking better decisions with respect to the stock management.

EOQ: Economic Order Quantity

The economic order quantity (EOQ) was introduced by Ford W. Harris in 1915. This concept calculates the request quantity which reduces to the minimum the inventory cost of the organization [10].

Considering an annual demand D, a cost of the order S, unity cost C, warehousing cost h, where the objective is to estimate the Q lot size, which minimizes the total of the annual cost. The EOQ [8, 11] equation is the following:

$$Q = \sqrt{\frac{2DS}{hC}} \tag{12.1}$$

The EOQ model is one of the most antique stock model into the operations field and inventory management. This model has been the main angle for many other models in these fields, partially due to the fact of its simplification to execute [12]. On the other hand, the EOQ has been criticized for the lacking response to the difficulties of the current business world [13].

Safety Stock

Generally, the organizations take decisions respect to their stocks based to their historical registers. These data are compiled through the time and are analyzed to determine possible inventory parameters, such as the safety stock. However, there is a possibility that data might own significant imprecision [14]. This type of situations puts in doubt the statistical traditional approach and is chosen for using the Cagan Safety Inventory [15]. Chopra defines the inventory as the stock that keeps in case the demand exceeds the expectations, it is useful to balance the uncertainty. The adequate safety stock level is based on three main factors: the demand, supply uncertainty, and the desirable product availability. As the demand or supply uncertainty increases, the required safety stock levels also expand [4].

Bayesians Networks

The Bayesians networks are probabilistic models used for the prediction and decision taking under uncertainty. The quantity required, the number of products

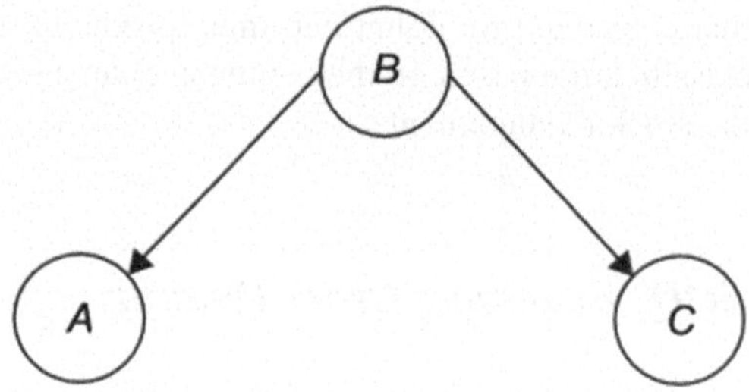

Fig. 12.1 Example of simple Bayesian network

to manufacture and the inventory have variations due to unexpected events. The Bayesian networks have been employed to estimate the probability of the offer and demand distribution, letting to make adjustments in the production and inventory plans [16]. A simple Bayesian network (Fig. 12.1) has random variables A, B, and C, which are the graph nodes. The artists represent the causal relations between them. The A and C variables have a causal dependency with variable B [17].

The Bayesians networks are adequate for the settlement of neuronal interactions due to their graphic nature and a rigorous subjacent theory. The edges of a Bayesian network are directional, for the trajectory neuronal signals proper modeling [16].

Fuzzy Logic

The fuzzy logic pretends to qualify imprecise information or with a certain vagueness grade, doing emphasis just in the fuzzy variables which represent this information and not the model as one [18, 19].

Uncertainty

Comes up from the lack of information (lexicon impression, something incomplete), particularly the inaccuracy of measurements [20].

Vagueness

The natural language vagueness is another aspect which limits the intention of being precise in an employed language to describe or sharing knowledge, communication, among others. The basic meanings of the words can be understood and have the capacity of communicate them in an acceptable rank, but generally, getting on the same level of understanding is not always possible [20].

The fuzzy logic allows to manage human subjectivities and system vagueness, perception, and reasoning. It is known that classic models define any systems or scenario mathematically and their results are rough values (crisp). In the other hand, the fuzzy logic handles variables that affect in the system's behavior obtaining more flexible outcomes. Therefore, with fuzzy logic, the variables are expressed in linguistic terms, for example: "large, medium and small," the relations are defined in rules based in If-Then terms, and the results are fuzzy subgroups, which can then be converted in rough values (crisp) using defuzzification techniques. The entry rough values are transformed and expressed in linguistic terms. In this step, the grade of membership of a variable is determined within a particular fuzzy class. This is specified across the evaluation of the membership function of the fuzzy group value [21]. A membership function defines the belonging point of a fuzzy group element mathematically. Then, the membership function defines the fuzzy subgroup A, where X is the domain of the variable in which A is sharped. In this approach, into the domain belonging to A, each element is defined with x values defining by steps for each X. Even though both are subsets, classic and fuzzy, are defined by the membership functions, the grade an element belongs to the classic subset is limited to be only zero or one. On the other hand, the value that each element takes is between the zero and one range, the above to define the fuzzy classification to which each element belongs [22]. This can be shown graphically in Fig. 12.2.

The helpfulness of the fuzzy sets lay in the capacity of modeling something uncertain or ambiguous. Some of the reasons why the fuzzy logic is employed to control processes are the next ones: when the process is too non-lineal or too complex to use conventional control designs. Another situation is because the fuzzy logic is less expensive, so they can obtain optimal results with data with a level of vagueness, considering the accuracy implies a cost on time, money, and resources, contributing to the expected utility just to a certain point, raising the costs of the model. By other hand, it is important to note that sometimes the organization is not able to provide information due to the lack of time or the risk of spread confidential information. In graphic of Fig. 12.3, it can be observed the precision cost of the information [23].

Exist two types of inference fuzzy systems: Mamdani and Sugeno. When the output of the membership functions are fuzzy groups, the inference fuzzy system Mamdani is the most common used [24]. The main idea of the Mamdani system is to describe the process between linguistic variables and to use the same ones as

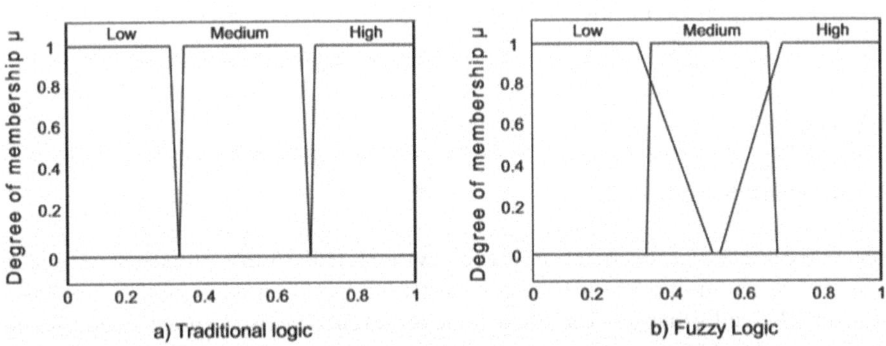

Fig. 12.2 Example of groups of (**a**) classical logic and (**b**) fuzzy logic

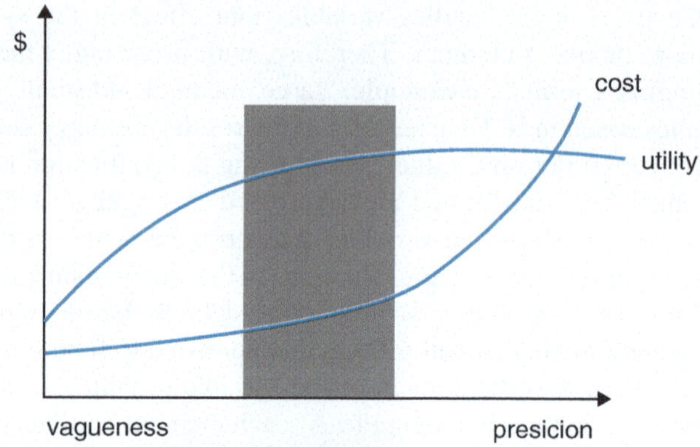

Fig. 12.3 Graphical cost benefit of accuracy of information

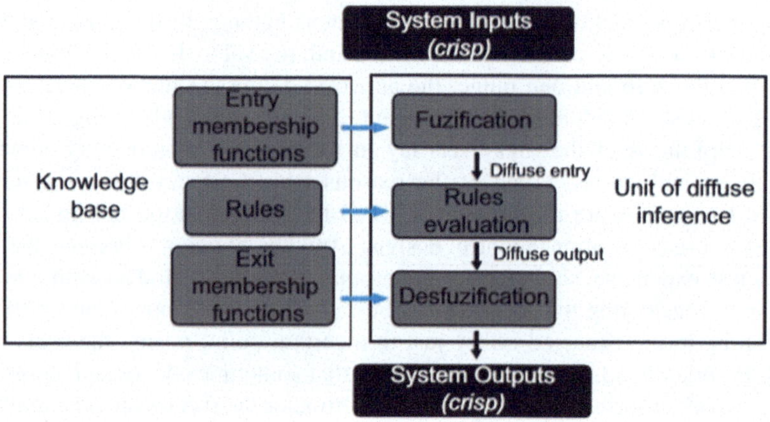

Fig. 12.4 Diffuse inferential system block diagram

entry data of the fuzzy control rules [25]. The Mamdani fuzzy inference system owns a knowledge basis, from where the entry membership functions are obtained, as the fuzzy rules and the output membership functions too. On the other side, also has a fuzzy inference unit, composed by the fuzzification system, the fuzzy rules evaluation, and the defuzzification process [26]. This can be shown in Fig. 12.4.

ANFIS

A Subsection Sample

The fuzzy logic and the fuzzy inference systems (FIS) were proposed by Zadeh [18], providing an option for the decision taking based on ambiguous, imprecise, or incomplete information. The fuzzy logic represents the used knowledge utilizing "If-Then" rules. FIS is mainly composed of fuzzy rules, membership functions, and defuzzification and fuzzification operations. Through the fuzzy inference, the tough entry data produce an ordinary output which is easy to understand and perform. There are three categories of fuzzy inference systems: Mamdani, Takagi-Sugeno (TS), and the Takagi-Sugeno Evolution (ETS). The identification of the basic rules is the key for a fuzzy inference system, presenting two main issues: (1) There are no standard methods for the human knowledge transformation or a basic rule experience. (2) Adjustments are required to the membership functions, to minimize the output mistakes, maximizing the performance. In the current project, ANFIS is used to achieve the FIS identification and its refinement.

Some ANFIS Uses

- The prediction of the daily precipitation using a variety of meteorological parameters of surface [27].
- In the econometric field, it is used to predict the performance of the stock market [28, 29].
- Breast cancer detection [30].

Sun et al. [31] studied the variable behavior of the demand of automobiles, how to forecast it and mention the tools to measure it. They compared and carried out the demand forecast of cars by analyzing the results of two models: Auto-regressive integrated moving average (ARIMA) combined with (TSDM) compared to a hybrid model of partial least square (PLS) regression and the adaptive network based on ANFIS giving the latter better results.

Candan et al. [32] focused on the study of the demand of products of the pharmaceutical industry, taking into account its complexity to be based on human factors such as seasonal and epidemic diseases and market shares of competing products among others. To make forecasts, they based on artificial neural network (ANN) topology with an ANFIS approach.

12.3 Methodology

Taking in consideration what was said in the previous sections, the next steps are established to describe the proposed tools for the problem solution.

First, it analyzes all the products of the company that make the inventory based on the information of the products that the organization provides, where the inventory is ordered using the ABC methodology. Subsequently, this methodology indicates a focus on the products within the classification type A. A product was selected from the best sellers and with high inventory turnover.

Then, the variables are modeled using fuzzy logic and experiments were carried out with the help of the ANFIS method using the Matlab tool. With this we can propose the necessary policies to support the decision maker. This can be seen in Fig. 12.5.

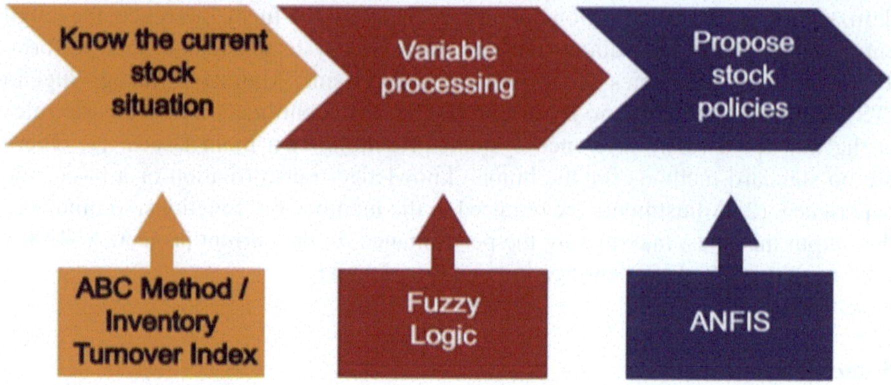

Fig. 12.5 Methodology

Applying ABC Method

The ABC method helps us to have a better inventory tracking, based on the sales and the volume of the products [6]. Approximately, the 20% of inventory provides to the economy the 80% of sales and incomes within the company, this is well-known as the "80–20" Rule. Stock is classified into three known groups as A, B, and C. The 20% of the articles represents to the A Group, the next 30% are articles from the B Group, and the last 50% is part of the C Group [7]. This allows us to focus our attention firstly in the products which contribute with higher income (A Group), taking care of the remaining classification based on the company possibilities, enabling a better decision take respecting to the inventory management.

Variable Processing

The human knowledge is represented generally by imprecision, vagueness, and approximates. In the real life, some definitions are handled with vagueness through words like: tall, medium, and short [33].

Fuzzy Logic

In literature, the theory provides a mechanism to represent linguistics constructions such as "many," "down", "medium," "often," "few." In general, the fuzzy logic offers an inference structure which allows proper capacities of human reasoning [20].

A fuzzy system is characterized due to the permission on using the experts' knowledge in an area they dominate, which works as a basis in the process automatization in such a way that, the knowledge is formalized, being ambiguous as a feasible algorithm formed by calculations like additions and comparisons [34].

Fuzzy Sets

Per its name, it is a group without a sharped outline, i.e., the transition since "belonging to the group" to "not belonging to the group" is gradual, being a soft transition, a characteristic of the membership functions which provide fuzzy and flexible sets along the modeling, where commonly linguistic expressions are used, e.g. "water is hot," or "the temperature is high" [8, 35].

Fuzzy Interface

It is known as the process of the mapping formulation, starting from an entry data to an exit using fuzzy logic. Mapping provides a basis from where the decisions can be taken, being judgment basis. The fuzzy inference process involves a variety of concepts such as membership functions, logic operations, and "if-then" rules. The main objective of the fuzzy inference process is to be formulated within expressions of natural [33] e.g.:

```
begin
{{antecedent} THEN conclusion {consequent}}
end
```

According to Chakraborty et al. [33], the different steps in the fuzzy inference process are as follows:

- Fuzzification of entry values: When the premise value is provided, such as an entry, this must correspond to one or more linguistic groups with some membership values, see Fig. 12.6.
- Strength calculation: After the entries are fuzzified, the grade for each part of the antecedents is satisfied for every known rule.
- The level of a rule is the strength rule of the correspondent one. i.e., if there is more than one antecedent, then the rule is calculated by the minimal standard operation.

There are different forms or graphics of the membership functions, such as triangular, trapezoidal, Gaussian, or with variations, presented in Fig. 12.7.

There are two types of fuzzy inference: Mamdani and Takagui-Sugeno, which differ in the output membership function, the one determined.

According to Jang and Sun [8, 35], the nature of the fuzzy models is similar to the "divide and conquer" meaning. The "if" (condition) "then" (action) rule divides the input space into many local fuzzy regions, while the consequent describes the behavior of a region through its constitution [36].

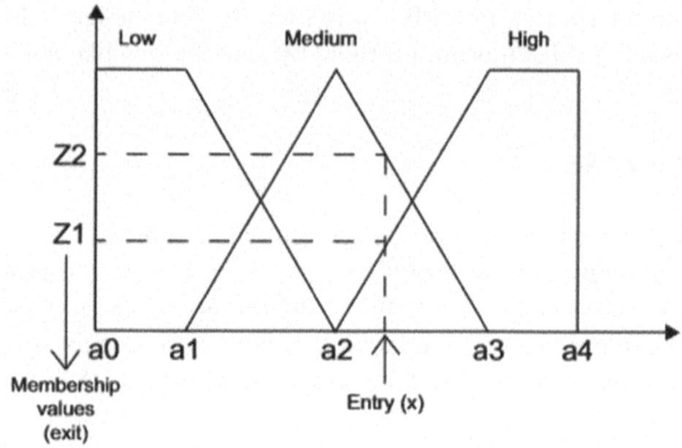

Fig. 12.6 Fuzzification

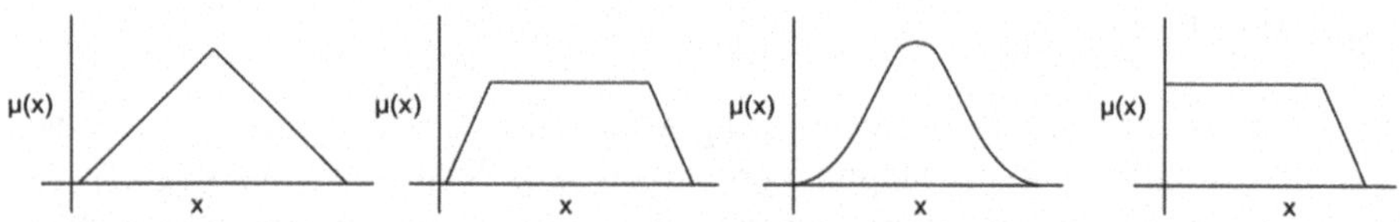

Fig. 12.7 Different forms of graphs corresponding to the membership functions

Defuzzification

Defuzzification means to convert a fuzzy value in a crisp one. This comes up due to the fuzzy results cannot be used as well in the applications. Therefore, it is mandatory to transform fuzzy quantities in crisp quantities or scalar ones to be processed in the future, through the defuzzification process. Defuzzification reduces the membership functions collections values in one quantity [37]. In the fuzzy control systems, the steps for the defuzzification involve the selection of one single value for the controller output. According to Rouhparvar and Panahi [38] we have many strategies.

Mamdani Model

To begin, we have the input of real data to the fuzzy logic system, whose values are numerical (crisp). The fuzzificator function or blurry convertor is to take the numeric values coming from the exterior, turning them in fuzzy values that can be processed by the inference mechanism. The fuzzy values are the input values membership levels to the different fuzzy conjuncts, on which the universe has been divided among the input variables. The objective of the inference system is to take the belonging levels supported by the rules, generating the output of the fuzzy system. The rules form the knowledge basis in which the fuzzy system keeps the linguistic system, allowing to solve the issue for which has been designed. A rule has two parts: the IF (antecedent) and the conclusion (Then). For example, if the output (inventory) is low, then the input (order size) is high. In a Mamdani system,

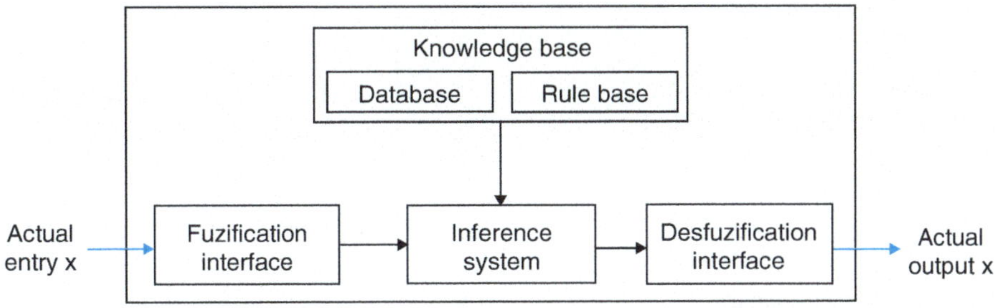

Fig. 12.8 General structure of a rule system based on the diffuse Mamdani type

Fig. 12.9 Centroid method

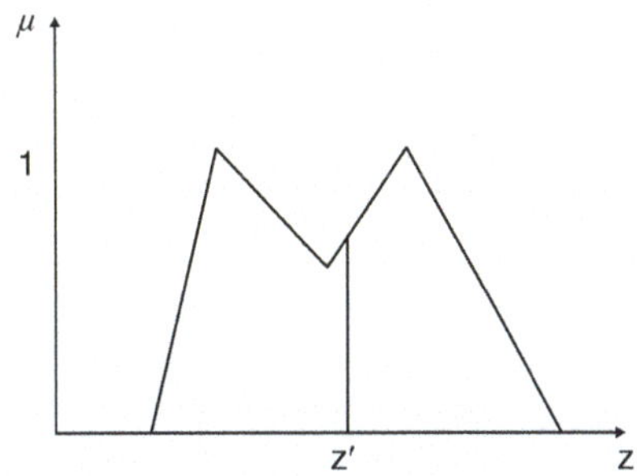

either the background and the rule conclusion is given by linguistic expressions. The output that is generated by the inference system is a fuzzy output that cannot be read by an external element. In order to read the output as a crisp value, the fuzzy output before must be processed by a defuzzification process, which gives the numeric output data. All this can be observed in Fig. 12.8.

- *Centroid Method*—This is the most used method. Also, named gravity center or area center [20]. It is determined by the area center of the belonging function [38]. Graphically it is perceived as in Fig. 12.9, where the result of the figure center from the operation is presented.

- *COS: Center of Sums*—The overlap zone is counted twice. This method compounds the consequential membership function, taking the algebraic sum of the outputs for each one of the fuzzy conjuncts that constitutes $\tilde{A}_1$, $\tilde{A}_2$, ... etc. The defuzzified value x^* is given by the Eq. (12.2). This method can be easily implemented by the user, being directed to the quick inference cycles. This method can be easily implemented by the user, being directed to the quick inference cycles.

$$x^* = \frac{\sum_{i=1}^{N} xi \sum_{k=1}^{N} \mu_{\tilde{A}k}\, x_i}{\sum_{i=1}^{N} \sum_{k=1}^{N} \mu_{\tilde{A}k}\, x_i} \tag{12.2}$$

- *MOM: Mean of Maximums: Center of Sums*—The rule mode consists on choosing the support value in which the membership function reaches the highest

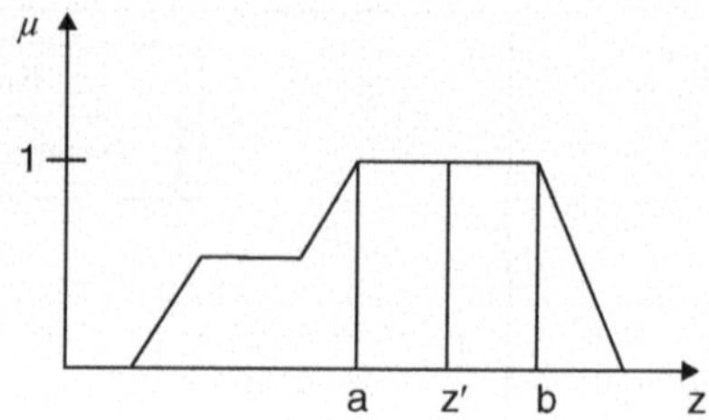

Fig. 12.10 MOM example

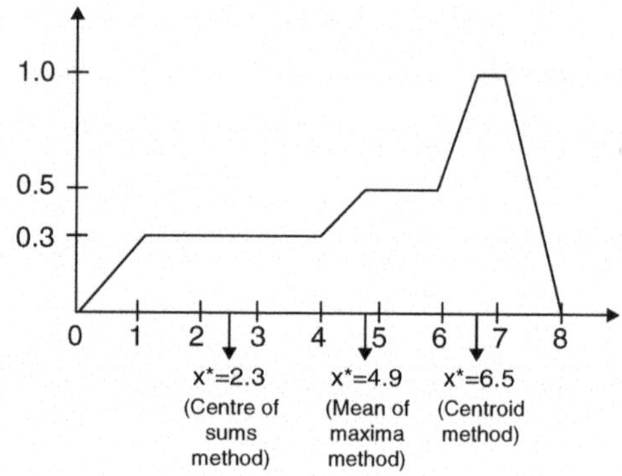

Fig. 12.11 Defined outputs of the three methods: Centroid, MOM, COS

peak. The mean maximum method takes a maximum affiliation [38]. The Eq. (12.3) shows that

$$x^* = \frac{\sum_{i=0}^{N} \sum_{i=0}^{N} \binom{n}{k} x x_{y^2} x_{y^2 1}{}^{n} Y \mu_{Ak^2}{}^{i} - x_i}{\sum_{i=1}^{N} \sum_{i=0}^{N} \binom{n}{k} x^k a^{n-k}} \tag{12.3}$$

The graphic remains as is shown in Fig. 12.10.

We can see graphically an example where the results can be appraised of each one of the methods previously mentioned. See Fig. 12.11.

Takagi-Sugeno Model

The Takagi and Sugeno model is based in the fuzzy logic system [18], a non-lineal system which assigns an input vector in a scalar output. A non-lineal Takagi-Sugeno forecaster uses a small window of past measurements to predict the next value in a time series [36].

In the Sugeno fuzzy system, the fuzzification mechanism and the fuzzy inference work in the same form as denominated Mamdani, but the based rules of the Sugeno knowledge are different, due to they have as output numeric values, hence does not require defuzzification. The Sugeno model employs the rules to produce an output in each rule. The output rules consist in a lineal combination of input variables and a constant terminus. The final output is the weighted weight of each output rule [39].

To calculate the fuzzy system output, the different consequences are weighted, keeping in mind the value in which the antecedent was activated for each one of the rules. The equation would be $w_1y_1 + w_2y_2 + \ldots \; w_ny_n \; w_1 + w_2 + \ldots \; w_n$

$$Y = \frac{w1y1 + w2y2 + \ldots wn\ yn}{wnyn\ w1 + w2 + \ldots wn} \tag{12.4}$$

Model Mamdani Vs. Takagi-Sugeno (TS)

The main difference between the methods Mamdani and Sugeno is found in the rules consequences. Mamdani utilizes fuzzy conjuncts as rules consequences while Sugeno employs lineal functions of the input variables as a rule consequence.

Some of the TS advantages are the following:

- Is computationally efficient.
- Works optimally with the lineal techniques (e.g. the PID control).
- Works great with the optimization and adaptation techniques.
- The output surface for the continuity has been guaranteed.
- Is adequate for the mathematical analysis.

The advantages of Mamdani are listed below:

- Is intuitive
- Has a wide acceptation
- Is adequate to handle human input sources

12.4 Inventory Policies Development

Rules

Based on the decision takers experience, a combination of different scenarios is created, specifying the expected result being observed through history. These fuzzy variables combination, from forecast, inventory, delivery time, and projects, provide us a total of 63 fuzzy rules that suggest us a result regarding the order tons to require to the foreign supplier. The generated rules from the process are exposed below (Table 12.1).

The following is an example of the "If-Then rules", where the use of the conditions (IF-AND) arises, which when they are met, give us a "THEN" response that affects our fuzzy output variable: "Order".

Rule 1: If {Forecast is Low} and {Inventory is Low}, Then {Order = Low}
Rule 2: If {Forecast is Low} and {Inventory is Medium}, Then {Order = Low}

Table 12.1 Fuzzy rules generated

Rules	Forecast	Inventory	Lead time	Project	Order
1	Low	Low	Low	Low	Low
2	Low	Low	Low	Medium	Low
3	Low	Low	Low	High	Low
4	Low	Low	Medium	Low	Low
5	Low	Low	High	Low	Medium
6	Low	Medium	Low	Low	Low
7	Low	High	Low	Low	Low
8	Low	Low	Medium	Medium	Low
9	Low	Medium	Medium	Low	Low
10	Low	Medium	Low	Medium	Low

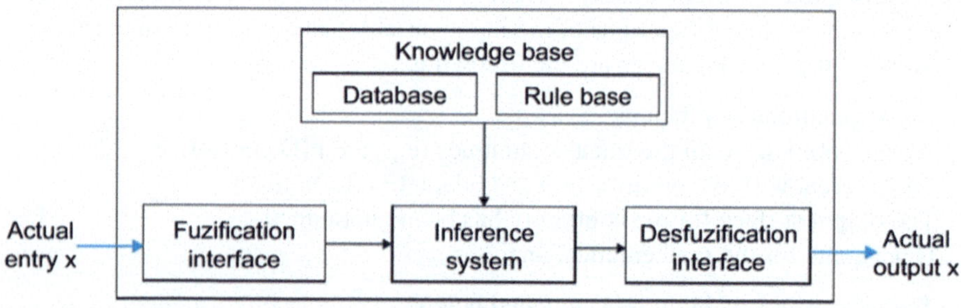

Fig. 12.12 Architecture of the proposed fuzzy inference model

The fuzzy model finally remains in the following way, as is shown on Fig. 12.12 where we can identify clearly the fuzzy variables involved in the case: forecast, inventory, delivery time, and projects; which are processed through the preset rules based on the Mamdani model, providing us in the order output variable result.

Output Defuzzification

Within the defuzzification process, it is required to pick the technique to implement this action. In this case, the centroid method was utilized, which is used commonly. The equation for discreet variables is listed in the Eq. (12.5).

$$\text{Centroid} = \frac{\sum_x \mu A(X)x}{\sum_x \mu A(X)} \tag{12.5}$$

This defuzzification method is also known as gravity center in literature. Consists in obtaining the area center under a formed curve by the fuzzy output function (see Fig. 12.13).

Fig. 12.13 Defuzzification method

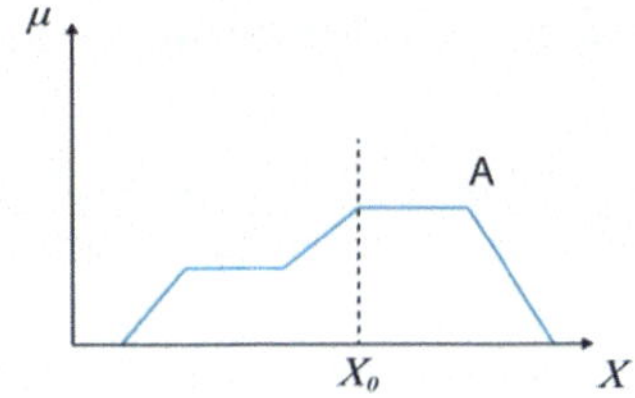

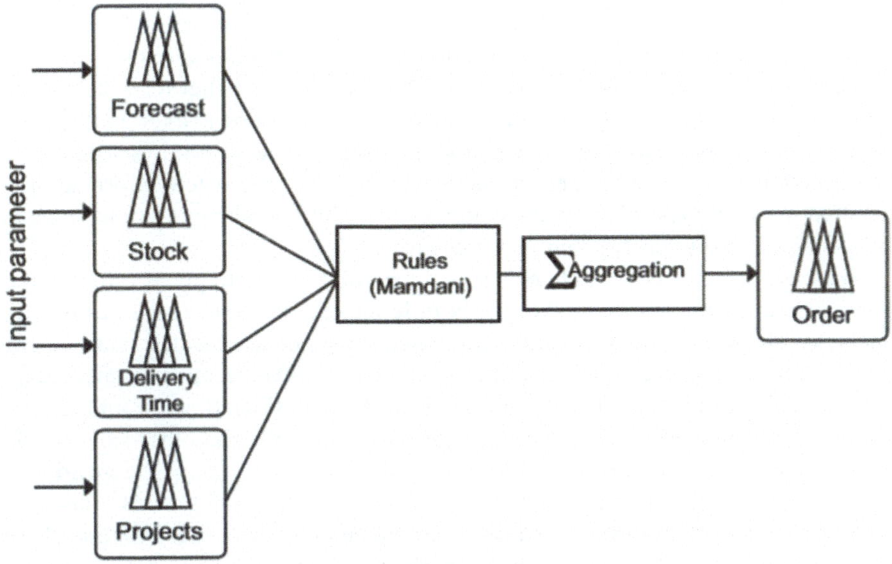

Fig. 12.14 ANFIS: Input variables with Mamdani rules

ANFIS

Keep in mind ANFIS is generated from a specific input or output, an inference system where the membership parameters are adjusted using the retro-propagation algorithm (artificial neuronal network), inducing the fuzzy system learns from the input or output data provided. The ANFIS model to employ is shown on Fig. 12.14, counting with.

- Input variables: They are 4: Forecast, inventory, delivery time, and projects.
- Belonging functions: We have three belonging groups in each one of the four variables: low, medium, and high.
- Rules: There are 63 fuzzy rules acquired from the experts experience combination.
- Input belonging function: One by each rule. Each value stills being fuzzified.
- Aggregation: Mamdani is used with the centroid method to defuzzify values.
- Output: Shows the process result: order variable, with a tons value to request (crisp).

12.5 Conclusion

In Table 12.2 it is shown the results that were obtained applying the proposed approach. It represents the six planning periods, where the company requires approximately 367.9 days in orders, which do not cover the demand completely, considering a 57 days of backlog. If the triangular ANFIS 1 process is utilized, after the six periods order there are 436.53 material days, which do not generate backlog, but a surplus of 11.67 days does. For the Gaussian ANFIS 2—an order of 452.4 days is shown, also without backlog, but also presenting a 27.49 days surplus.

The use of ANFIS in this case indicates that the result is better than the current procedure, avoiding backlog generated due to the growth in the market. In this case, we note situations where few information is possessed and is not exact, this type of methodology is effective and provides results. Also, when few information is required, it is not expensive for the company regarding on obtaining time, human, and economic resources for their procurement.

The case presents certain characteristics regarding to the required information. Some of the main features are: relatively little information, one-year sales record, one-year inventory, and 6-month item purchase process. Besides, part of the information is mandatory (Delivery time and part of stock) presenting an uncertainty and vagueness of this one. If only the safety stock is considered, we would require a huge stock number, considering the high uncertainty. On the other side, we do not count with enough information to modify the forecast. Some of the mentioned methodologies in the chapter handle the uncertainty, such as safety stock, Bayesian networks, forming part of the possibilities theory, among others. Nevertheless, these topics which handle the probability theory do not include the vagueness, resulting of subjective causes, as it would be "the delivery time is high,", or "the stock is low.". This class of situations is common in problems where the experts take their decisions based on non-quantitative situations, in where we can apply the ANFIS methodology, which embraces the fuzzy logic.

For future research, the following is suggested:

- A study to build a more robust multi-objective method, that is, in addition to considering sales, take into account the other factors that can be considered for the formation of each of the groups indicated by the ABC method.
- Include the remaining products of groups A and B, to improve customer achievement, with the aim of reducing delays.

Table 12.2 Results of proposed approach

Concept	1	2	3	4	5	6	Total	Difference
Business order	51.53	61.02	61.02	40.80	51.00	102.00	367.90	
Business-stock-out	14.00	0.00	10.00	18.00	15.00	0.00	57.00	424.90
Triangular ANFIS1	70.31	76.53	65.00	75.41	72.76	76.53	436.53	11.67
Gaussian ANFIS2	72.04	76.50	77.10	75.30	74.90	452.4	27.49	

References

1. B. Samanta, S.A. Al-Araimi, An inventory control model using fuzzy logic. Int. J. Prod. Econ. **73**(3), 217–226 (Elseiver, 2001). https://doi.org/10.1016/S0925-5273(00)00185-7
2. C.A. Garcia, A. Ibeas, R. Vilanova, A switched control strategy for inventory control of the supply chain. J. Process Cont. **23**(6), 868–880 (Elseiver, 2013). https://doi.org/10.1016/j.jprocont.2013.04.005
3. U.E. Kocamaza, H. Taşkınb, Y. Uyaroğlu, A. Göksu, Control and synchronization of chaotic supply chains using intelligent approaches. Comput. Ind. Eng. **102**, 476–487 (Elseiver, 2016). https://doi.org/10.1016/j.cie.2016.03.014
4. S. Chopra, P. Meindl, *Supply Chain Management: Strategy, Planning, and Operation*, 5th edn. (Pearson, Prentice Hall, New Jersey, 2013)
5. A.K. Chakravarty, Multi-item inventory aggregation into groups. J. Oper. Res. Soc. **32**(1), 19–26 (JSTOR, 1981). https://doi.org/10.2307/2581465
6. M.A. Millstein, L. Yang, H. Li, Optimizing ABC inventory grouping decisions. Int. J. Prod. Econ. **148**, 71–80 (Elseiver, 2014). https://doi.org/10.1016/j.ijpe.2013.11.007
7. B.E. Flores, D. Clay Whybark, Multiple criteria ABC analysis. International Int. J. Oper. Prod. Manag. **6**(3), 38–46 (MCB UP Ltd, 1986). https://doi.org/10.1108/eb054765
8. H. Altay Guvenir, E. Erelb, Multicriteria inventory classification using a genetic algorithm. Eur. J. Oper. Res. **105**(1), 29–37 (Elseiver, 1998). https://doi.org/10.1016/S0377-2217(97)00039-8
9. V. Pareto, *Manual of Political Economy: A Critical and Varioum Edition* (Oxford University Press, Oxford, 1971)
10. D. Simchi-Levi, P. Kaminsky, E. Simchi-Levi, *Designing and Managing the Supply Chain: Concepts, Strategies, and Case Studies*, 3rd edn. (McGraw-Hill/Irwin, New York, 2008), p. 498
11. S. Chopra, P. Meindl, *Supply Chain Management: Strategy, Planning, and Operation*, 2nd edn. (Pearson, Prentice Hall, New Jersey, 2007)
12. L.E. Cárdenas-Barrón, G. Treviño-Garza, H. Wee, A simple and better algorithm to solve the vendor managed inventory control system of multi-product multi-constraint economic order quantity model. Expert Syst. Appl. **39**(3), 3888–3895 (Elseiver, 2012) https://doi.org/10.1016/j.eswa.2011.09.057
13. C.H. Glock, K. Schwindl, M.Y. Jaber, An EOQ model with fuzzy demand and learning in fuzziness. Int. J. Serv. Oper. Manag. **12**(1), 90–100 (Inderscience, 2012). https://doi.org/10.1504/IJSOM.2012.046675
14. A. Kumar, P.T. Evers, Setting safety stock based on imprecise records. Int. J. Prod. Econ. **169**, 68–75 (Elsevier, 2015). https://doi.org/10.1016/j.ijpe.2015.07.018
15. M. Cagan, When to record transactions, Netplaces, http://www.netplaces.com/accounting/keeping-track-of-transactions/when-to-recordtransactions.htm
16. J. Shin, S. Kim, J.-M. Lee, Production and inventory control of auto parts based on predicted probabilistic distribution of inventory. Digit. Commun. Netw. **1**(4), 292–301 (Elseiver, 2015). https://doi.org/10.1016/j.dcan.2015.10.002
17. J. Reguero-Alvarez, J. Diaz-Garcia, Aplicacion de las redes bayesianas dinamicas a la prediccion de series de datos y a la deteccion de anomalias, Tesis de Maestria, Departamento de Ingenieria Informatica, Universidad Autonoma de Madrid, Madrid, 2011
18. L.A. Zadeh, Fuzzy sets. Inform. Cont. **8**(3), 338–338 (Science Direct, 1965)
19. L.A. Zadeh, Is there a need for fuzzy logic? Inf. Sci. **178**(13), 2751–2779 (ScienceDirect, 2008)
20. S.N. Sivanandam, S. Sumathi, S.N. Deepa, *Introduction to Fuzzy Logic Using MATLAB* (Springer, Berlin, Heidelberg, New York, 2007). https://doi.org/10.1007/978-3-540-35781-0
21. N. Alavi, Quality determination of Mozafati dates using Mamdani fuzzy inference system. J. Saudi Soc. Agric. Sci. **12**, 137–142 (Elseiver, 2013). https://doi.org/10.1016/j.jssas.2012.10.001

22. N. Alavi, V. Nozari, S.M. Mazloumzadeh, H. Nezamabadi-pour, Irrigation water quality evaluation using adaptive network- based fuzzy inference system. Paddy Water Environ. **8**(3), 259–266 (Springer, 2010). https://doi.org/10.1007/s10333-010-0206-6
23. J. Yen, R. Langari, *Fuzzy Logic: Intelligence, Control, and Information* (Prentice Hall, Upper Saddle River, 1999)
24. S.M. Mazloumzadeh, M. Shamsi, H. Nezamabadi-pour, Evaluation of general-purpose lifters for the date harvest industry based on a fuzzy inference system. Comput. Electron. Agric. **60**(1), 60–66 (ScienceDirect, 2008). https://doi.org/10.1016/j.compag.2007.06.005
25. E.H. Mamdani, S. Assilian, An experiment in linguistic synthesis with a fuzzy logic controller. Int. J. Man Mach. Stud. **7**(1), 1–13 (ScienceDirect, 1975). https://doi.org/10.1016/S0020-7373(75)80002-2
26. R. Mirabbasi, S.M. Mazloumzadeh, M.B. Rahnama, Evaluation of irrigation water quality using fuzzy logic. Res. J. Environ. Sci. **2**(5), 340–352 (Academics Journals Inc., 2008). https://doi.org/10.3923/rjes.2008.340.352
27. E. Aldrian, Y.S. Djamil, Application of multivariate ANFIS for daily rainfall prediction: influences of training data size. Makara J. Sci. **12**(1), 7–14 (2008). https://doi.org/10.7454/mss.v12i1.320
28. M.A. Boyacioglua, D. Avcib, An Adaptive Network-Based Fuzzy Inference System (ANFIS) for the prediction of stock market return: The case of the Istanbul Stock Exchange. Expert Syst. Appl. Int. J. **37**(12), 7908–7912 (Elseiver, 2010). https://doi.org/10.1016/j.eswa.2010.04.045
29. T. Ansari, M. Kumar, A. Shukla, J. Dhar, R. Tiwari, Sequential combination of statistics, econometrics and Adaptive Neural-Fuzzy Interface for stock market prediction. Expert Syst. Appl. **37**(7), 5116–5125 (ScienceDirect, 2010). https://doi.org/10.1016/j.eswa.2009.12.083
30. D. Nawgaje, R.D. Kanphade, Implementation of ANFIS for breast cancer detection using TMS320C6713 DSP, in *Proceedings on International Conference and workshop on Emerging Trends in Technology (ICWET)*, No. 13, pp. 8–11 (2011)
31. B. Sun, B. Li, G. Li, K. Zhang, Automobile demand forecasting: An integrated model of PLS regression and ANFIS. Int. J. Adv. Inform. Sci. Serv. Sci. **5**(8), 429–436 (2013). https://doi.org/10.4156/aiss.vol5.issue8.52
32. G. Candan, M.F. Taskin, H.R. Yazgan, Demand forecasting in pharmaceutical industry using artificial intelligence: neuro-fuzzy approach. J. Mil. Inform. Sci. **2**(2), 41–49 (Sakarya Universitesi, 2014). ISSN: 2148-3124
33. N. Chakraborty, S. Mondal, M. Maiti, A deteriorating multi-item inventory model with price discount and variable demands via fuzzy logic under resource constraints. Comput. Ind. Eng. **66**(4), 976–987 (Elseiver, 2013). https://doi.org/10.1016/j.cie.2013.08.018
34. B. Martín del Brio, A. Sanz Molina, *Redes neuronales y sistemas difusos 3er edicion* (RA-MA EDITORIAL, Mexico, 2006). ISBN: 978-84-7897-743-7
35. J.S.R. Jang, C.-T. Sun, E. Mizutani, *Neuro-Fuzzy and Soft Computing: A Computational Approach to Learning and Machine Intelligence* (Prentice Hall, New Jersey, 1997). ISBN:0-13-261066-3
36. C. Pereira da Veigaa, C.R. Pereira da Veigaa, W. Puchalskic, L. dos Santos Coelhoc, U. Tortatoa, Demand forecasting based on natural computing approaches applied to the foodstuff retail segment. J. Retail. Cons. Serv. **31**, 174–181 (Elseiver, 2016). https://doi.org/10.1016/j.jretconser.2016.03.008
37. S. Rajashekaran, G.A. Vijayalksmi, *Neural Networks, Fuzzy Logic and Genetic Algorithms: Synthesis and Applications* (PHL Learning Private Limited, New Dehli, 2013). ISBN: 978-81-203-2186-1
38. H. Rouhparvar, A. Panahi, A new definition for defuzzification of generalized fuzzy numbers and its application. Appl. Soft Comput. J. **30**, 577–584 (Elseiver, 2015). https://doi.org/10.1016/j.asoc.2015.01.053
39. A. Dwivedi, M. Niranjan, K. Sahu, A business intelligence technique for forecasting the automobile sales using Adaptive Intelligent Systems (ANFIS and ANN). Int. J. Comput. Appl. **74**(9), 7–13 (2013). https://doi.org/10.5120/12911-9383

Part IV
Renewable Energies

Chapter 13
Development of Renewable Energy Resources in Global Energy Interconnection

Thu Yein Min, He Haiyang, Michael G. Tyagunov, Tatiana A. Shestopalova, and Aung Ko

Contents

13.1 Introduction 175
13.2 GEI-Backbone of Decarburization 177
13.3 European Energy Ring 180
13.4 Asian Energy Ring 181
13.5 Development of Renewable Energy Sources by Gobitec and Asia Network Projects in Northeast Asia 182
13.6 Two Concepts for the Construction of the Asian Energy Ring 182
13.7 Power Plants in the Unified Power System of the Russian Federation 183
13.8 Unified Energy System (UES) of Russia 184
13.9 Optimization Model of Power System with Installations Based on Renewable Energy Sources (Fig. 13.11) 184
13.10 Development of Renewable Energy in Myanmar 184
13.11 Conclusion 189
References 190

13.1 Introduction

When technology develops today, energy efficiency requirements are increasing every year. Almost all developed countries develop and carry out some renewable energy projects. The main principle of using RES is to get renewable energy from our environment and use it for electricity. Currently, using RES has the good level, not only related to human development and energy security, but also by economic feasibility. Renewable energy has significant advantages from the view point of social and ecology significance. Hydro energy is an important part of RES, it has sustained and high-efficiency characteristics. However, the utilization of hydro energy is a complex task. To create a hydroelectric power station, special information, mathematical and software tools are needed for the feasibility study of the projected hydroelectric power station in countries, where market relations

T. Y. Min (✉) · H. Haiyang · M. G. Tyagunov · T. A. Shestopalova · A. Ko
National Research University "MPEI", Moscow, Russia

P. Vasant et al. (eds.), *Data Analysis and Optimization for Engineering and Computing Problems*, EAI/Springer Innovations in Communication and Computing,
https://doi.org/10.1007/978-3-030-48149-0_13

Fig. 13.1 GEI is a clean energy-dominant, electric-centric modern energy system that is globally interconnected, jointly constructed and mutually beneficial to all. It is an important platform for large-scale development, transmission and consumption of clean energy resources worldwide. In essence, GEI is "Smart Grid + UHV Grid + Clean Energy"

are only at the stage of their formation. In this regard, there are reliable data on resource consumption and optimization for energy parameters and operating types of hydroelectric power plants to improve efficiency to utilize renewable energy resources, and to select the composition of generating stations based on RES and their optimal parameters. By global experience, just utilization of RES in energy supply systems for consumers does not always provide reliable and uninterrupted power supply due to the physical characteristics of RES. In this regard, the energy consumer in Myanmar is best organized by sharing RES such as water, solar, and wind. To improve efficiency to operate hydroelectric power stations, there is need to build a set of program, which can do optimization for design parameters and operating types of all components of hydro-electrical station in accordance with the specified optimality criterion [1].

Global Energy Interconnection (GEI)

$$\text{GEI} = \text{Smart Grid} + \text{Ultra} - \text{High Voltage electric grid} + \text{Clean Energy}.$$

Over the past 25 years, the global energy landscape has undergone significant changes, and many greater changes will be possible in the future. It provides an analysis of long-term energy forecasts among government, intergovernmental, and private organizations using a unique methodology. These forecasts agree that in the absence of an ambitious climate policy, world energy consumption will grow by 20–30% or more by 2040 and beyond, mainly due to the use of fossil fuels. This growth is driven by population growth and economic growth in the global "East," while energy consumption in the global "West" remains roughly unchanged [2] (Figs. 13.1 and 13.2).

China announced a global initiative for the development of a Global Energy Interconnection (GEI). The conclusions of the initial feasibility study were presented

Fig. 13.2 By 2050, GEI will basically come into being

by Chinese President Xi Jinping, on September 26, 2015, at the United Nations Sustainable Development Summit.

The research was carried out by the State Grid, which is the largest public utility. It proposed the goal of meeting world energy needs, development in environmental-friendly ways, and finding alternative plans [3].

To further study and solve global energy problems, a new non-profit international organization-Global Energy Interconnection Development and Cooperation Organization (GEIDCO) was launched in March 2016, Beijing, China. The purpose of GEIDCO is meeting the global power demand with clean and green alternatives. At present, the number of GEIDCO members has expanded to 756 from 106 countries. Divided by continents as follows: Asia-492 members, Africa-96 members, Europe-71 members, Oceania-10 members, North America-20 members, and South America-67 members. The number of GEIDCO members is still growing [4].

13.2 GEI-Backbone of Decarburization

To solve the challenges for sustainable energy development, GEIDCO proposed a plan, integrated by the smart grid, UHV (ultra-high voltage), and clean energy. The aim of GEIDCO is to promote the energy supply by using electricity instead of other kinds of energy, and it will significantly reduce carbon dioxide emissions and promote decarburization.

GEI realizes energy transmission by focusing on electricity, and the power grid is the main manifestation. At present, the power grid faces three main challenges. Firstly, the demand for energy and power in large-scale load centers is growing rapidly; Secondly, the proportion of random and intermittent renewable energy sources such as wind and solar power in the power grid is increasing; Thirdly, the requirements for safe and stable operation of the power grid are increasing.

These issues facing the power grid have also made it more urgent to build a globally connected power grid.

Under the precise control of the smart grid and high capacity of the UHV transmission lines, the electricity system could storage much more capacity of renewable energy resources, such as hydro, wind, solar, etc. By using more electricity from RES, the consumption of fossil energy will gradually decrease.

The GEIDCO Roadmap to 2035

Before 2035, global power flows were dominated by cross-regional and transnational power exchanges within continents, and transcontinental power exchanges were in their infancy.

In 2035, the total scale of global intercontinental and interregional power flow will reach 280 million kW, of which 50 million kW will be transcontinental power flows. Global installed capacity will reach 15.7 billion kW, with 66% of clean energy installed capacity, including 21% of wind power, and 22% of solar power [5].

The GEIDCO Roadmap to 2050

By 2050, the clean energy base will enter a large-scale development stage, forming a global clean energy optimization configuration, multi-energy complementarily, and cross-time and mutual-benefit patterns.

In 2050, the total scale of global intercontinental and interregional power flows will reach 720 million kW, of which transcontinental power flows will reach 130 million kW. In 2050, global installed capacity will reach 26.7 billion kW, with clean energy installed capacity accounting for 83%, of which wind power is 25%, and solar power is 41%.

In the long run, the Arctic region has become an important global clean energy base. Through the large-scale development of wind energy resources in the Arctic region and solar energy resources in the equatorial region, it can meet the growing power demand in Asia, Europe, and North America, and the power flow across continents will further increase [5] (Fig. 13.3).

China's Energy Plan to Solve Global Problems

To achieve the goal of a global deployment that is based on inter-time zone, off-season, and large-scale use of clean energy, depending on resource allocation, electricity needs, and climate and environmental management needs, Chinese

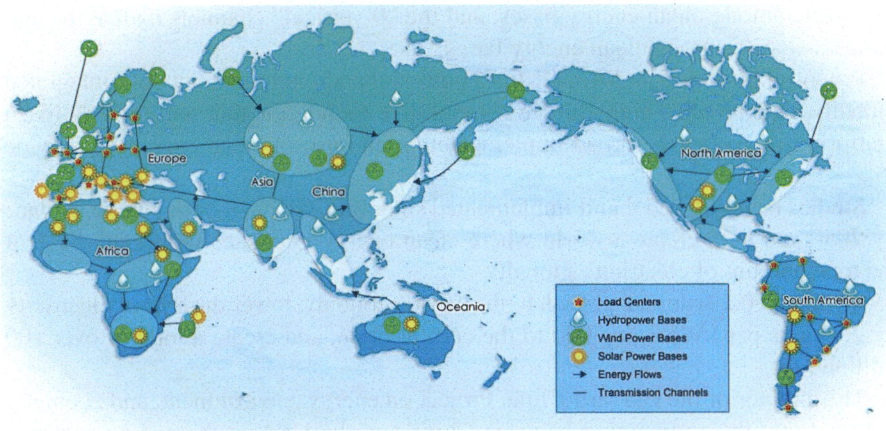

Fig. 13.3 Global energy interconnection of the world

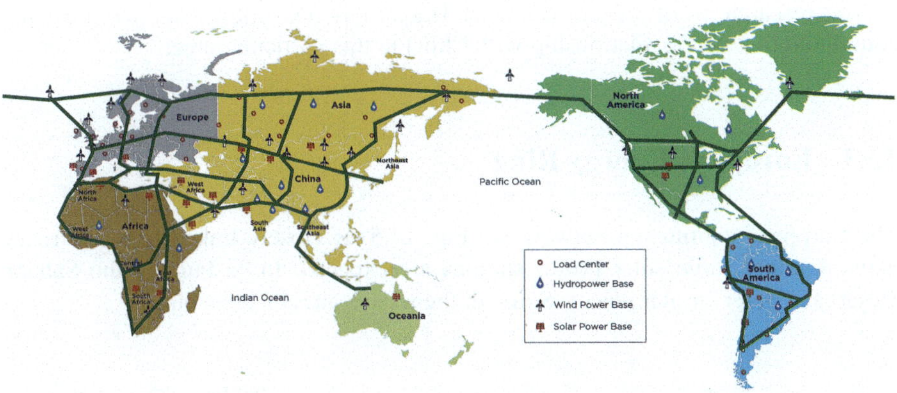

Fig. 13.4 9 horizontal and 9 vertical channels of GEI-backbone grid in long term

experts have proposed a plan that is based on building a national backbone network and a multinational network to build "9 horizontal and 9 vertical" backbone energy networks, an extensive combination of large clean energy bases and load centers, as shown in Fig. 13.4.

"9 horizontal and 9 vertical" backbone network frame includes:

1. Asia-Europe-Africa "4 horizontal and 6 vertical" interconnected channels;
2. The "4 horizontal and 3 vertical" interconnection channels of the Americas;
3. The Arctic energy interconnection.

The backbone network realizes the power transmission from the clean energy base to the load center, while the "9 horizontal" channels realize the aim of cross-time zone among clean energy bases, and the "9 vertical" channels realize the aim of cross-season among clean energy bases.

Proposed by China in 2015, GEI is a powerful and intelligent network integrated globally, with an ultra-high voltage network like foundation that will provide like a platform to develop the broad implementation and use of clean energy worldwide [6].

Studies have showed that if implemented smoothly, GEI will basically be in place by 2050 and help create a world where clean energy generates more than 81% of the total amount of electricity globally.

It will add 0.2% annual growth to the world economy, lower the cost of electricity by 2.8 cents per kWh compared to the current price, and create a total of over 100 million jobs.

The director of the Harvard-China Project on energy, environment, and economy believed that the cooperation between China and the USA in the energy sector is very important and there are a number of ways of sustaining such cooperation within the GEI framework.

The research work we are doing at Harvard is also important, we are very committed to a strong relationship with China in this particular area.

13.3 European Energy Ring

The European continental network (in Fig. 13.5) is closely linked to the African network, since large solar power stations are expected to be built in the Sahara Desert and other similar places. Some of them have already been built.

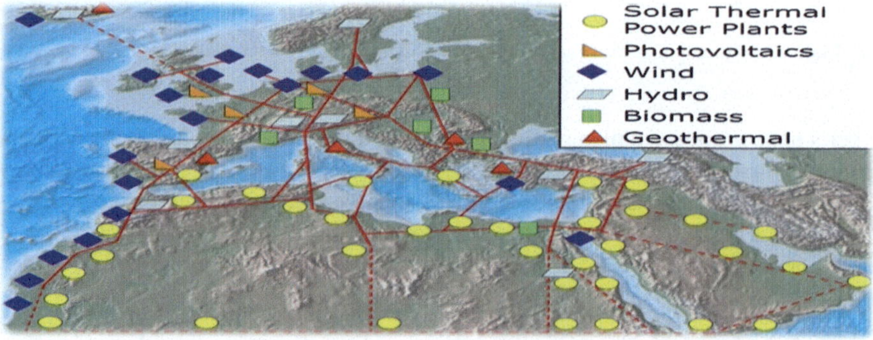

Fig. 13.5 European continental network

13.4 Asian Energy Ring

The Asian energy ring should link Russia's hydroelectric power stations and solar power plants in the Gobi desert with Japan. The transmission line will pass through China and the two Koreas. In Kazakhstan and Central Asia, there is a project to connect solar power plants to this energy ring.

Connecting the complexity of the network systems of Russia, Mongolia, China, Japan, and South Korea seems like a good idea for BP CEO Robert Dudley, who spoke exclusively to RT at the forum (Fig. 13.6).

When they talk about the economic efficiency of this project, they operate on several things. Firstly, it is mutual redundancy of power systems. But the current standards in the Russian Federation are so high compared with the world level that in some additional redundancy from external power systems it is not necessary to speak. Secondly, interchange of electricity. But Russian energy tariffs are uncompetitive with Chinese ones (Fig. 13.7).

Japanese law prohibits the import of electricity. It turns out that the initiators of the Asian Energy Ring project in Japan should achieve the lifting of this ban.

The large Asian energy ring should cover North and Southeast Asia. In the center of generation solar stations are located in the Gobi desert. But there are political obstacles in this project. Japan, for example, prohibits the import of electricity by law. Therefore, the project will depend on political decisions.

Fig. 13.6 Possible move of the northern ring

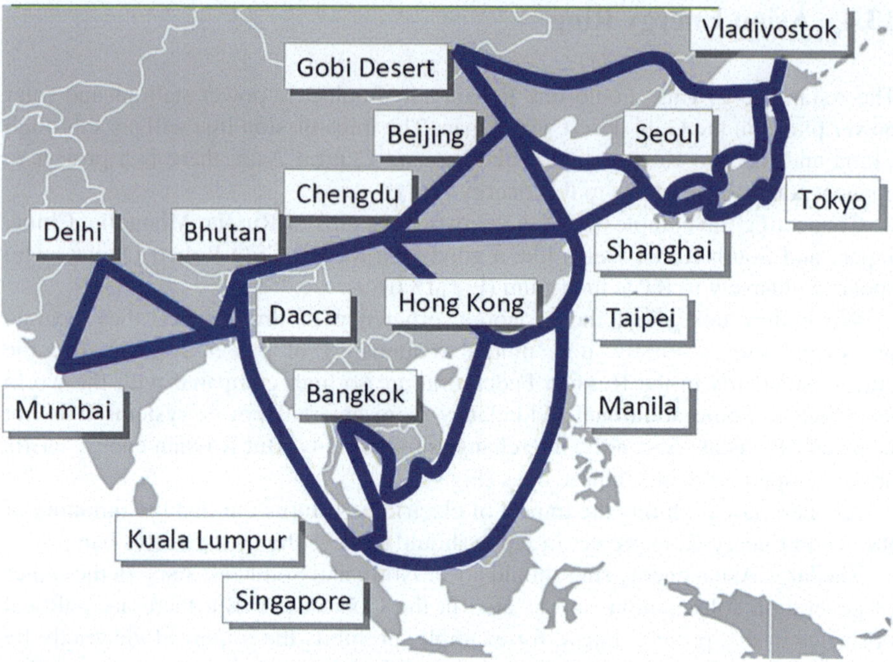

Fig. 13.7 State grid of China—Asia super network with partners

13.5 Development of Renewable Energy Sources by Gobitec and Asia Network Projects in Northeast Asia

The Power Charter provides unilateral rules for energy cooperation [7]. It is very important in the international efforts to create a legal framework for the Gobitec and the Asia Super Network with RES for the energy security of countries based on the rules of accessible, aggressive markets, and sustainable development in Northeast Asia (Fig. 13.8).

The Energy Charter plays an important role in creating the Asian Super Network with renewable energy sources. It should ensure the energy security of countries, which are based on the rules of fully competitive markets and long-term sustainable development [8].

13.6 Two Concepts for the Construction of the Asian Energy Ring

Constructing power stations in Russia, Kazakhstan, and Mongolia, and then transfer the electricity generated by these countries to China, Korea, and Japan.

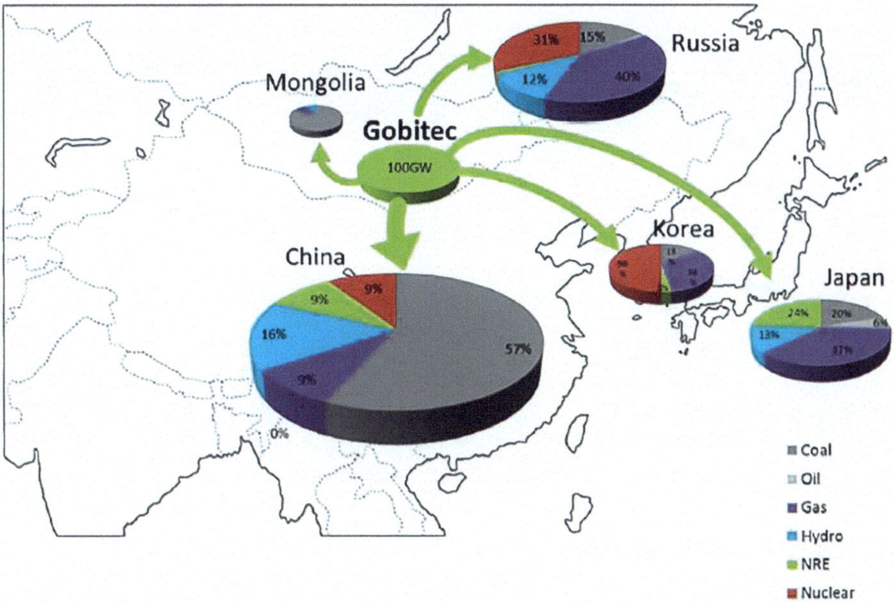

Fig. 13.8 Generation mix is based on 2030 BAU forecast. Japan : Scenario of phasing-out nuclear by 2030

There are two concepts for the development of the Euro-Asian super network:

1. A network for transferring working capacity from regions where energy resources are available to areas where energy consumers are located.
2. A network of communication between self-supplied energy regions, balancing excess needs for electrical energy and heat due to power flows.

13.7 Power Plants in the Unified Power System of the Russian Federation

The absorption of huge power-plants nearby locations not far from primary energy sources (PES) raises their efficiency, when transfer costs of fuel reduce from production sites to the place of use (Fig. 13.9).

For the right choice of a development path, it is proposed to use the experience of the Unified Energy System of Russia.

The second important principle: the generation structure ensures the reliability and survivability of the Unified Energy System.

Fig. 13.9 Russia energy distributed system

13.8 Unified Energy System (UES) of Russia

Partition of a huge electric-system that is complicated with networks, generators, and consumers, complete into smaller components. Manageability of UES was ensured by a single ownership and a vertically integrated system of operational dispatch and technological management (Fig. 13.10).

Controllability of the Unified Electric System was ensured by a single ownership right and a vertically integrated system of operational dispatch and technological management [9].

13.9 Optimization Model of Power System with Installations Based on Renewable Energy Sources (Fig. 13.11)

For each part of the distributed power system with installations on the baseline of RES, a model of the optimal structure of generating and consuming installations is constructed [9].

13.10 Development of Renewable Energy in Myanmar

At present, wind and solar energies are RES that can get more than other resources in Myanmar [10]. According to modern data, the gross resources of Myanmar's solar and wind energy are 1.15 million TWh and 1.82 thousand TWh (at an altitude

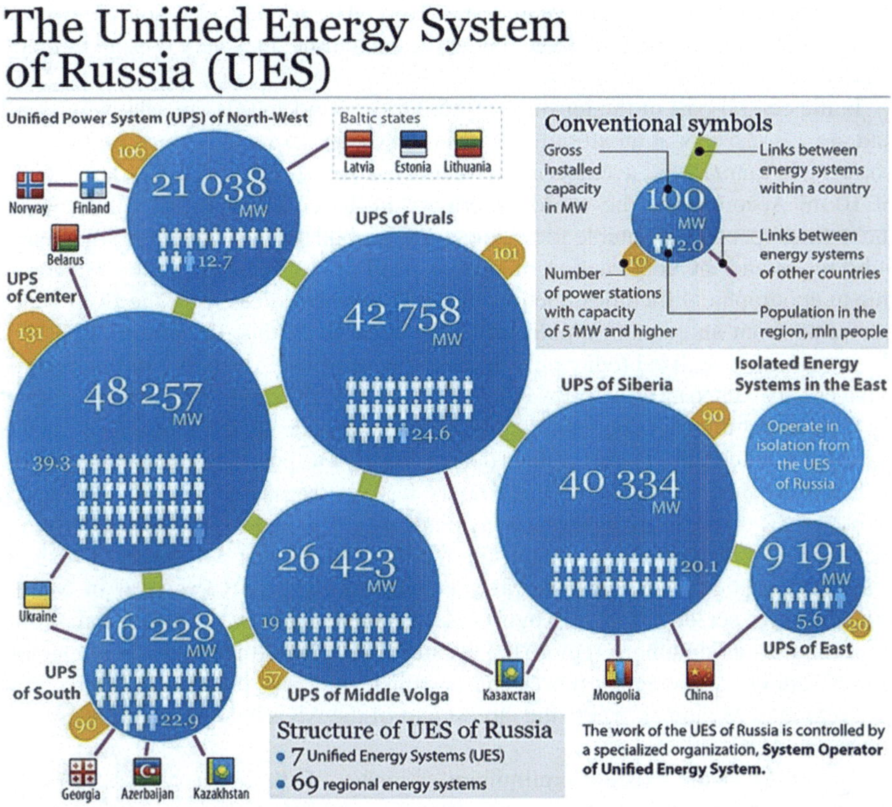

Fig. 13.10 Unified energy system of Russia

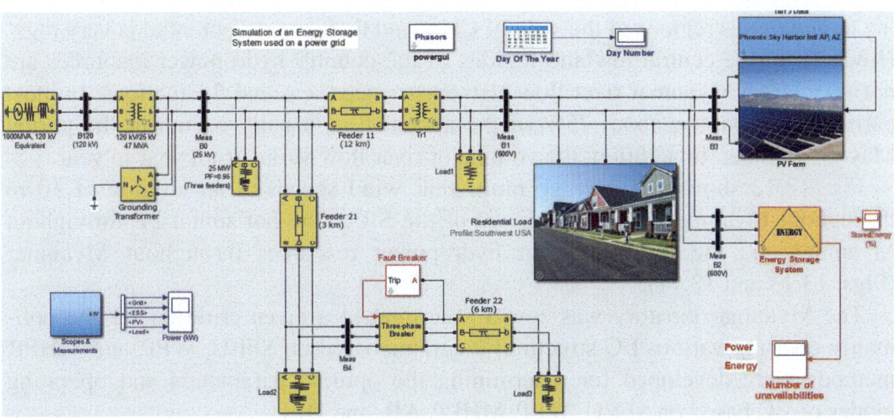

Fig. 13.11 Model of distributed power system based on RES

of 10 m above sea level) per year, respectively. The resources of solar and wind energy of Myanmar are quite large, but at the same time, they vary considerably by region.

In the central part of the territory of Myanmar, wind speeds are relatively small and are 1.2–2.5 m/s at an altitude of 10 m. However, in some regions on the west coast of Myanmar, the average multi-year wind speeds reach 5.5 m/s at a height of 10 m. According to the world experience in the development of wind energy, these wind speeds are suitable for efficient use for both energy supply of emergency substations and the construction of system wind power stations (WPP). However, due to geographic difficulties, the power supply of these areas from the ECO is not advisable from an economic point of view. Because of this, the power supply of these regions is realized today only at the expense of DEU or BEU. In this regard, wind energy can be an effective way to energy supply numerous AP in these regions.

Solar radiation (SR) depends mainly on the latitude of the terrain, i.e. at the equator, it takes the greatest value, decreasing to the poles. Myanmar is located near the equator, and therefore the country has vast resources of AOC. The daily arrival of SR on the horizontal receiving platform (PP) ranges from 3.5 to 6 Wh/m^2, depending on the latitude of the terrain and the time of year. CP levels also show seasonal variations. For example, at a latitude of 21°, the CP arrival in March is 6 kWh/m^2 per day, and in August—4.13 kWh/m^2 per day. In addition to the northernmost mountainous region, the use of solar cells for the provision of electric power supply to emergency power can be considered very promising. The central vast region of Myanmar, where the largest part of the AP is located, is the best place to use the AOC.

According to the existing preliminary estimates of the Ministry of Electricity, the value of Myanmar's hydropower potential is 108,000 MW. Despite the unique potential, hydropower Myanmar today is used only by 3%. Hydropower resources are unevenly distributed throughout the country. There are a significant number of rivers in the northern and eastern mountainous regions of Myanmar, as well as in the mountainous regions of the state of Chin, and their energy potential is very high. However, in the central lowland regions of the country hydropower resources are not very large. Myanmar river flow also experiences seasonal fluctuations. In most parts of the country, about 75% of the annual flow usually occurs in the period May–September. In addition, the volume of river flow varies from year to year.

Fig. 13.12 shows the average multi-year wind speeds at an altitude of 10 m throughout Myanmar, the daily arrival of the SR at the horizontal PP throughout Myanmar and the distribution of hydropower resources throughout Myanmar (Figs. 13.13 and 13.14).

The Myanmar territory was zoned according to a given criterion for the optimality of using various EC structural diagrams based on SFEU, WPP, and MHPP, methods were developed for determining the optimal parameters and operating modes of EC based on SFEU, WPP, MHPP, AB, and DEU.

To successfully solve the problems posed in this article, the authors developed and used the appropriate application program "STAD" developed in C ++ high-level programming language and with the help of AP research the integrated use

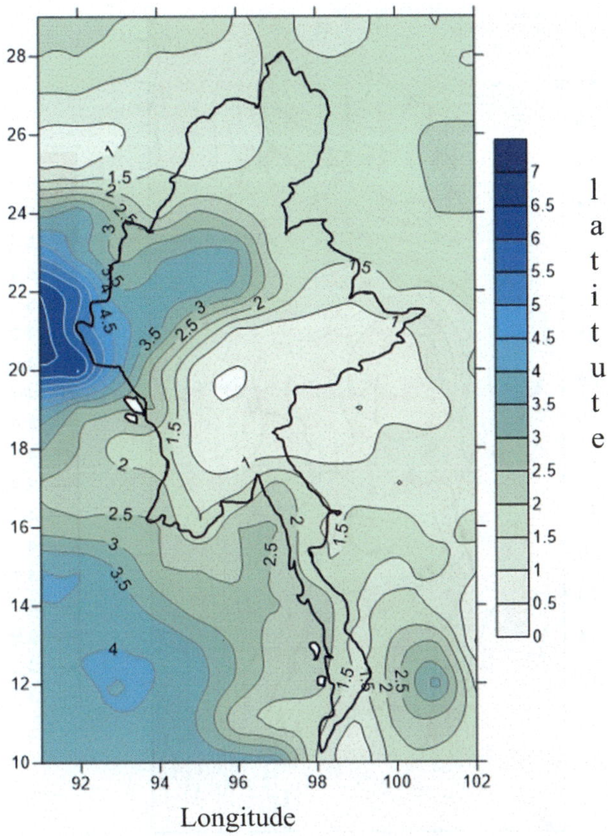

Wind energy resources at an altitude of 10 m - 1.93 thousand TWh/year in Myanmar.

Fig. 13.12 Regular wind speeds (m/s) at 10 m height across the territory of Myanmar

of HEC to solve the problem of electricity for many autonomous rural consumers in Myanmar [11]. The AP "STAD" is designed for the calculation of three types of HEC.1st: heat pump (GSHP), solar energy (PV), accumulator battery (AB), the diesel electrical generator (DGEN) (HEC1); 2nd: air source heat pump (ASHP), PV, AB. and DGEN (HEC2); 3rd: air-conditioning with boiler, PV, AB, and DGEN (HEC3). With the help of AP, estimated the economic efficiency of these HEC at different costs of the components of the HEC and comparison of different components. AP "STAD" simulates the physical attitude of the energy system and its cost over period of exploitation, including the costs of the installation and further operation. AP "STAD" allows the project engineer to compare a variety of different variants for the design of the power system and determine its economic and technical advantages associated with variability of weather conditions. Structure of AP "STAD" contains three main algorithms: algorithm for heating load calculation, algorithm for solar

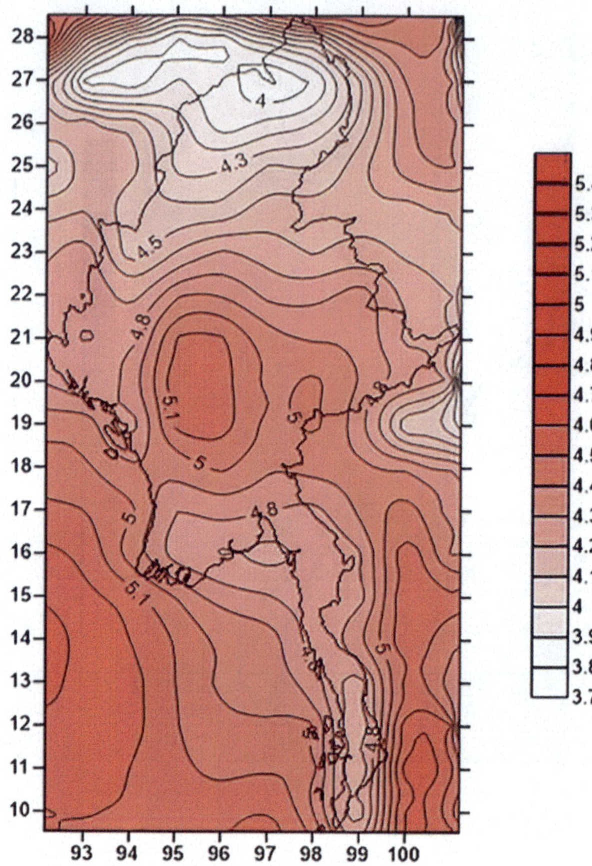

Fig. 13.13 Daily arrival of the SR at the horizontal reception point in Myanmar (kWh/m^2)

radiation calculation on arbitrary-oriented solar modules, and algorithm for the energy balance calculating for three types of HEC, taking into account different kinds of constraints such as equality and inequality and isoperimetric conditions. AP "STAD" presents the results of calculations of optimization in the form of text file, which facilitates comparison of various results and allows to determine the economic and technical advantages of various HEC [1].

In the future work, energy complex of renewable energy resources for electricity supplement to self-governing consumer of Myanmar will be developed in according to take account of many criteria [10]. In this regard, firstly will be need to collect geo-information data of project location, information support of wind and solar data, definitions of various type of criteria, and methods for solar and wind calculations.

For the first time for the conditions of Myanmar, a general task has been set for the mathematical substantiation of primary energy parameters and modes of the power system of numerous agricultural consumers in this considered region of Myanmar. The accepted criteria of optimality for solving the set general problem are substantiated, the basic calculation equations for the electric power balance

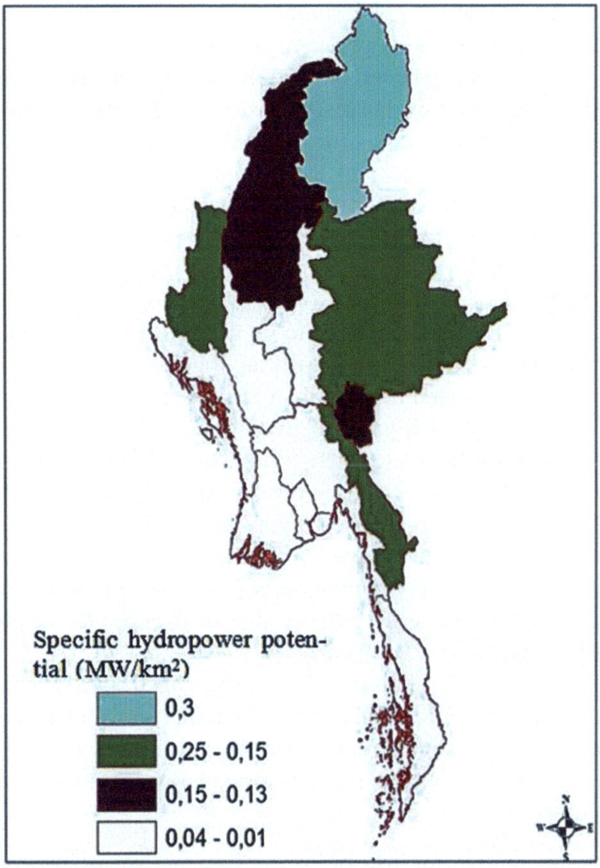

Fig. 13.14 Distribution of hydropower resources in Myanmar

are compiled taking into account the limitations of the operating modes of the EC components and the permissible operating conditions of the EC components. Typical electrical load plots of autonomous rural power consumers for minimum, average, and maximum levels of power consumption have been developed [10].

13.11 Conclusion

The energy demand will be increasing day by day in the future in the world. At present all the countries are facing with the deficit of electricity and need to develop in renewable sector. To obtain the efficiency, stability, and safety of power supply the countries are trying to interconnect the distributed energy network.

To construct the project of the interconnecting network, it is needed to combine the technologies of ultra-high voltage grid, smart grid, and power sources based on renewable energy resources. The costs of these projects are very high and all countries need to cooperate to develop GEI.

Global energy interconnection is a great project not only to solve the problem of shortage of energy, but also to instead of the other kinds of energies to power energy. Power energy is more efficient, stable, and environment-friendly.

With development of technologies and sciences, the power-generation cost of RES is cheaper and cheaper, the electricity based on RES is more competitive than traditional power generation. By integrating RES into the grid system, smart grid technology provides us a way to achieve the target of clean air and energy independence [12].

In 2016 China proposed to set up the organization GEIDCO for developing GEI. Until now organization is working together with more than 600 other organizations which are located in more than 85 countries.

Currently, in Myanmar, hydropower plays a large role in the fuel and energy sector of the country, where more than 60% of the capacity and power generation is provided by hydroelectric power plants. According to the latest data from the RSM Ministry of Energy, 8 new hydropower plants with a capacity of 1691 MW and 3 solar energy station have already been developed with a total capacity of 470 MW. According to the Government's plan, over the next 10 years, power plants based on renewable energy will be able to occupy 9% (without hydropower plants) of Myanmar's generating capacity.

In the works of other graduate students from Myanmar it is shown that in coastal regions it is more efficient to use wind power plants instead of solar energy station, and in mountainous regions (in the north and east of the country) it is effective to use small hydropower stations instead of solar energy station to supply the country's numerous rural populations.

References

1. T.Y. Min, M.G. Tyagunov, G.M. Deriugina, *Evaluation of the Effectiveness of the Use of Programs in the Design of Power Complexes Based on Renewable Energy Resources* (IGI Global, 2020)
2. Resources for the Future, www.rff.org
3. CTC Global, www.ctcglobal.com
4. Global Energy Interconnection Development and Cooperation Organization, https://en.geidco.org/overview/brochure/
5. L. Jun, S. Fulong, Y. Xiaoxiao, Research on global energy interconnection backbone grid planning. J. Glob. Energy Interconnect. **1**(5), 527–536 (2018)
6. J.S. Feng, Q.L. Zhao, Environmental benefit assessment under the vision of global energy interconnection, in *IOP Conference Series: Earth and Environmental Science*, 2018
7. R. Kreindler, Are tribunals setting new limits on access to international jurisdiction? ICSID Rev. **5**(1), 37–43 (2010)
8. R. Leal-Arcas, *Solutions for Sustainability* (Springer, 2019)
9. M. Tyagunov, Distributed energy system's is the future of the world's power industry, in *2017 2nd International Conference on the Applications of Information Technology in Developing Renewable Energy Processes & Systems (IT-DREPS)*, 2017

10. M.G. Tyagunov, T.Y. Min, Analysis of ways of solving the problem of hybrid energy complexes based on reserve for power supply of autonomous rural consumers in Myanmar, in *2018 Renewable Energies, Power Systems & Green Inclusive Economy (REPS-GIE)*, Casablanca, Morocco, 23–24 Apr 2018
11. University of Petra, lms.uop.edu.jo
12. *XXth International Multidisciplinary Scientific Geo Conference, Surveying, Geology and Mining, Ecology and Management – SGEM 2020*, 27 June–06 July 2020, www.sgem.org

Part V
Information Technologies

Chapter 14
Main Metric Components in the Generation of Mixed Indicators: An Application of SGVD Methodology

Roman Rodriguez-Aguilar

Contents

14.1 Introduction 195
14.2 Generalization of the PCA with Metrics 196
14.3 Mixed PCA Application 200
14.4 Conclusions 205
References 206

14.1 Introduction

Principal component analysis (PCA) is one of the multivariate analysis methodologies widely used as the first step in the analysis of a data set. The main utility of PCA is that it is a statistical method that allows reducing the size of a correlated data set, keeping as much information as possible from the original data.

PCA is a methodology used for exploratory data analysis and in unsupervised models since the objective is to construct a k-dimensional space of principal component variables from an n-dimensional data set (with $k < n$). Therefore, PCA allows condensing information provided by multiple variables in a set of main component variables, which in some cases for a better graphic representation it is convenient to remain with 2 or 3 (main component variables) dimensions. Therefore, PCA is a very useful method in its application in various areas related to the analysis

The original version of this chapter was revised: Editor "Roman Rodriguez-Aguilar" is affiliation has been updated. The correction to this chapter is available at https://doi.org/10.1007/978-3-030-48149-0_21

R. Rodriguez-Aguilar (✉)
Facultad de Ciencias Económicas y Empresariales, Universidad Panamericana, Ciudad de México, México
e-mail: rrodrigueza@up.edu.mx

P. Vasant et al. (eds.), *Data Analysis and Optimization for Engineering and Computing Problems*, EAI/Springer Innovations in Communication and Computing,
https://doi.org/10.1007/978-3-030-48149-0_14

of large correlated data sets. Similarly, PCA is a technique that can be used as part of the exploratory data analysis, prior to the realization of supervised or unsupervised support models.

It is important to mention that the PCA can be performed from two perspectives, standard PCA and multiple correspondence analysis (MCA for its acronym in English), the difference between both methods lies in the type of variables that are considered, in the case of standard PCA it works only with quantitative variables and in the case of MCA only with categorical variables. But the objective is the same, to reduce the size of the data and concentrate the information on a smaller set of new principal component variables. Depending on the data that will be used, standard PCA or MCA is applied, segmenting the analysis into qualitative and quantitative approaches. However, it is of great interest in some PCA applications to decrease the size of a mixed data set, combining qualitative and quantitative characteristics. Therefore, in recent years the use of mixed PCA has been extended, a methodology that allows the mixing of qualitative and quantitative variables.

14.2 Generalization of the PCA with Metrics

Standard PCA

A common problem in data analysis is that a large number of variables are handled, some of which may be statistically related (correlated) and, therefore, would be redundant. If n measurements are being made for each of m experimental variables, m vectors (or series of data), each of n components, are generated in m-dimensional space. In this space, each coordinate axis represents a specific variable that is being measured. The redundancy and error (or noise) associated with the measurement process imply that the distribution of data on the m-dimensional space can have a complicated form in which it is difficult to detect useful patterns.

The statistical analysis of the data is simplified and clarified if the redundancy between the different variables is eliminated, which can be achieved by "rotating" the coordinate system of the m-dimensional space mentioned so that, in the new coordinate system, the most relevant statistical information is manifested only through some of the new main component variables generated. The series of data to be subjected to a principal component analysis, or PCA, is usually subtracted from the value of their respective means so that the input data has a 0 average. The dispersion matrix, or covariance matrix, generated from these new data is a symmetric matrix in which the main diagonal represents the variances of each series of measurements, while the elements outside the main diagonal represent the covariance between different measurements. These covariances are a measure of the redundancy between the data series. When two series of data (both with 0 means) are statistically independent, we expect their covariance to be close to 0 in magnitude. Therefore, a coordinate system in which the covariance matrix is a diagonal matrix is a system in which the statistical dependence between the variables has been eliminated by rotating the coordinate system to generate a diagonal covariance matrix, the new variables provide more accurate information about the statistical properties of the data.

The standard PCA is based on two fundamental concepts of linear algebra, eigenvalues, and eigenvectors (sometimes also called characteristic values and vectors or, less frequently, eigenvalues and eigenvectors). In a very descriptive way, the concept of eigenvectors could be understood as the result of the multiplication of a matrix A and a vector $\overline{v}$, where the vector resulting from said multiplication is a multiple of the original vector, that is $A\overline{X} = k\overline{X}$, where k is the eigenvalue of the matrix A corresponding to the eigenvector $\overline{X}$. There are some relevant properties for eigenvectors that should be highlighted [1]:

1. It is only possible to obtain the eigenvectors of a square matrix (but not for all square matrices).
2. The eigenvectors of a symmetric matrix are orthogonal to each other (which generates uncorrelated principal component variables).
3. By scaling an eigenvector and multiplying it by the associated matrix, a multiple of the same eigenvector is obtained (therefore it is possible to standardize them). From the above, it is inferred that the product of any eigenvector of a matrix by a nonzero constant is also an eigenvector of the same matrix.

By multiplying a matrix by one of its eigenvectors, a multiple of the original vector is obtained, the value by which the eigenvector is multiplied is known as an eigenvalue. Therefore, an eigenvalue corresponds to every eigenvector and an eigenvector corresponds to every eigenvalue. An important result of linear algebra (see, for example, [2]) shows that symmetric matrices S, such as the covariance matrix, can be diagonalized in the sense that they can be expressed in the form $S = P$ ^ 1 DP, where D it is a diagonal matrix and P is an invertible matrix that, in the case of symmetric matrices, turns out to be an orthogonal matrix. A diagonal matrix is one in which all the components outside the main diagonal are 0. An orthogonal matrix is made up of lines and columns that form a base of orthonormal vectors. That is, the lines that constitute an orthogonal matrix are unit vectors that are orthogonal or "perpendicular" to each other; if the point product of a line of an orthogonal matrix is taken with any other line, the result will be 0. The same applies to the columns. It is important to mention that, in the diagonal matrix generated in this process, the elements of the main diagonal are precisely the eigenvalues of the original symmetric matrix S. The order of the eigenvalues in the diagonal matrix can vary according to the selection of the matrices orthogonal, so the representation is not unique. However, in the PCA method, this order is selected in such a way that the highest eigenvalues appear first, corresponding to the new principal component variables with greater variance and, presumably, greater statistical interest.

The fundamental idea of the PCA method is to generate, by means of orthogonal diagonalization, a set of linear combinations of the original variables (with 0 means) such that the covariance matrix is diagonal. In the standard PCA method, the main component variables generated correspond to an eigenvector, and the relevance of

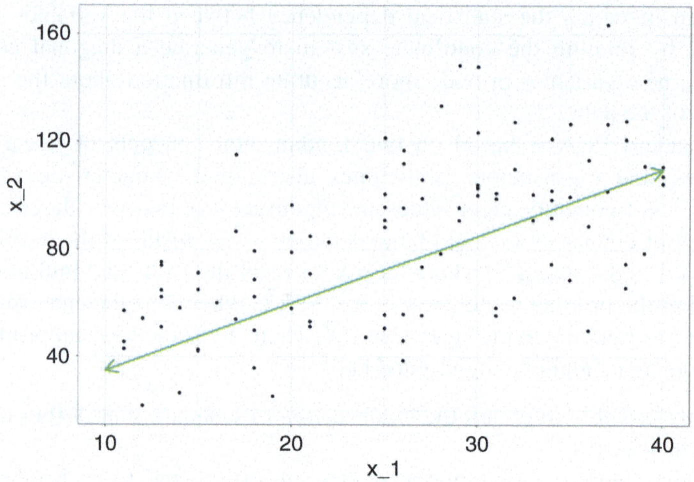

Fig. 14.1 Projection of space ($n = 2$) in the first component ($K = 1$)

the components is established in decreasing order according to the corresponding eigenvalue (the proportion of variance explained).

In a two-dimensional space, the interpretation of the first main component would correspond to the direction of the vector that represents the greatest variance of the original data. The projection of each observation with respect to this vector corresponds to the value of the first component. For this observation, the linear combination of the original variables makes up the first main component (Fig. 14.1).

Given the orthogonality property of the eigenvectors, the second component would correspond to the direction of the vector that represents the greatest variance without being correlated with the first component. Therefore the value of each main component C_i is obtained as the linear combination of the original variables. The main component variables can be understood as a result of a weighted sum of the original variables, these weights correspond to the eigenvectors.

$$C_i = \sum\nolimits_{ip} \theta_{ip} X_p \tag{14.1}$$

If the linear combination is normalized it implies that the sum of all weights is one.

$$\sum\nolimits_{i}^{p} {\theta^2}_{ip} = 1 \tag{14.2}$$

The values of θ correspond to the loadings that correspond to the weight of each component and represent the importance of each original variable in each principal component variable generated.

Generalized Singular Value Decomposition

The PCA metrics is a general version of the standard PCA previously described, the metrics that are used correspond to weights applied to the each row or column of the data matrix. This allows combining PCA and MCA analysis for mixed data. For this, generalized singular decomposition (GVSD) of a Z matrix [3, 4] is used.

$$Z = U \Lambda V^T \tag{14.3}$$

where Z is a real dimension matrix $n\ x\ p$; N is the diagonal matrix of the weights of the n observations; M is the diagonal matrix of the weights of the p columns; Λ is a diagonal matrix $r\ x\ r = \left(\sqrt{\lambda_1}, \sqrt{\lambda_2}, \sqrt{\lambda_3} \ldots, \sqrt{\lambda_r}\right)$ of the eigenvalues of ZMZ^TN and Z^TNZM, where r is the range of Z; U is the matrix $n\ x\ r$ of the (r) eigenvectors of ZMZ^TN for which $U^TNU = I_r$, where I_r is the identity matrix of dimension r;V is the matrix $p\ x\ r$ of the (r) eigenvectors of Z^TNZM for which $V^TMV = I_r$.

Where the main components are obtained from the resolution of an optimization problem, denoted as follows:

Max$\|ZMV_i\|^2N$, subject to

$$V_i{}^T M V_j = 0 \forall 1 \leq j \leq i,$$

$$V_i{}^T M V_i = 1 \tag{14.4}$$

where v_r solutions are the eigenvectors of Z^TNZM.

From the GSVD application, it is extracted that, regardless of the data used in the Z matrix, through the use of metrics by observations and by variables (N and M), it is possible to obtain as a particular case PCA and MCA. The matrix Z is the result of processing the original data matrix (X), matrix (N) and (M) [5].

In the case of PCA, the data to be used corresponds to n observations with p variables, using a X data matrix of dimension $n\ x\ p$. The data processing corresponds to centering and normalizing the data to construct the matrix Z (correlation matrix $\frac{1}{n}Z^T Z$). The n observations will be weighted by $\frac{1}{n}$ and the p columns are weighted by 1. Therefore, $N = \frac{1}{n}I_n\ I\_n$ and $M = I_p$.

For the MCA, the data to be analyzed correspond to n observations and p categorical variables in a matrix X of dimension $n\ x\ p$. Each categorical variable has j levels. In order to process the data, it is necessary to code each level of each categorical variable as a binary variable defining the indicator matrix G. To generate the ACM, correspondence analysis will be applied to this indicator matrix, where the component variables of the rows (observations) and the columns (levels) are obtained by PCA to two matrices, the profiles of rows and columns. In this case, the matrix Z will be the matrix G previously constructed, where the rows (observations) are weighted by $\frac{1}{n}$ and the columns (levels) are weighted by $\frac{n}{n_s}$ where n_s are the number of observations that belong to that level. So the matrix $N = \frac{1}{n}I_n$ and the matrix $M = \text{diag}\left(\frac{n}{n_s}, s = 1, \ldots, m\right)$.

14.3 Mixed PCA Application

Even when it is a methodology of recent implementation, there is a set of applications in different statistical packages. In the case of *R*, Ade4 [6], FactoMineR [7], and recently PCAmixdata [8] have been implemented, which is based on PCA with metrics using GSVD, including Standard PCA and MCA as particular cases.

An application of PCA with mixed data will be developed in the next section. A database corresponding to a set of qualitative and quantitative variables on Mexican households will be analyzed, belonging to the National Survey of Household Income and Expenditure (ENIGH by the acronym in Spanish) 2016 conducted by the National Institute of Statistics and Geography [9]. The variables considered are a sample of 500 observations and 12 variables of the ENIGH and correspond to sociodemographic characteristics of households, as well as variables related to the pattern of household consumption (Table 14.1).

Eigenvalues

When estimating the eigenvalues, we will obtain as many eigenvalues as the number of variables considered in the analysis. The sum of the eigenvalues is equal to the original information (total inertia) $p1 + m - p2$ and the first two dimensions recover 18% of the total inertia (Table 14.2).

Graphical representation of the results. Observation graph using the first and second components as axes (Fig. 14.2).

Figure 14.2 shows the map of main components where the observations are represented on a two-dimensional map corresponding to the first and second main components.

The level map for categorical variables shows the first two components as axes and identifies the levels of qualitative variables. The level map shows the relationship of each categorical variable with the first two main components. For example, in those cases where the education of the head of household corresponds to higher or postgraduate education, it is related to higher values in components one and two. On the other hand, those heads of households without education concentrate higher values in component 2 and lower values in component 1 (Fig. 14.3).

Correlation map uses the first two components as axes and shows their correlation with the quantitative variables. Correlation map indicates that the age of the head of

Table 14.1 Variables to consider from the ENIGH 2016

Variable name	Description	Type
Household class (clase_hog)	Differentiation of households based on the type of consanguineous, legal, affinity or custom relationship between the boss and the other household members	Qualitative Value Label 1 Unipersonal 2 Nuclear 3 Expanded 4 Compound 5 Co-resident
Sex of the head of the household (sexo_jefe)	A biological distinction that classifies the head of the household as male or female	Qualitative Value Tag 1 Male 2 Female
Age of the head of the household (edad_jefe)	Years between the date of birth of the head of the household and the date of the interview	Qualitative
Formal education of the head of the household (educa_jefe)	Formal education of the head of the household	Qualitative Value Label 01 Without instruction 02 Preschool 03 Incomplete primary 04 Complete primary 05 Incomplete secondary 06 Complete secondary 07 Incomplete high school 08 Complete high school 09 Incomplete professional 10 Complete professional 11 Postgraduate
Income from work (ingtrab)	Sum of income earned by work, like otros_trab. subordinate, as independent and from other jobs	Quantitative
Consumption of food outside the home (ali_fuera)	Expenditure on food and beverages consumed outside the home	Quantitative
Spending on tobacco consumption (tobacco)	Spending on tobacco	Quantitative
Housing expenditure (vivienda)	Housing expenditure, conservation services, electricity, and fuels	Quantitative
Energy expenditure (energía)	Expenditure on electricity and fuels	Quantitative
Health expenditure (salud)	Expenditure on health care, outpatient care, hospitalization, and medicines	Quantitative
Spending on personal accessories (acces_pers)	Spending on accessories and personal effects	Quantitative
Savings deposits (deposito)	Deposits in savings accounts, batches, savings banks, etc.	Quantitative

Source: Own elaboration

Table 14.2 Eigenvalues from PCAMix

Dimension	Eigenvalue	Proportion	Cumulative
dim 1	2.3116592	10.507542	10.50754
dim 2	1.7619767	8.008985	18.51653
dim 3	1.3657946	6.208157	24.72468
dim 4	1.2509182	5.685992	30.41068
dim 5	1.2348986	5.613176	36.02385
dim 6	1.2114162	5.506437	41.53029
dim 7	1.1579577	5.263444	46.79373
dim 8	1.0896466	4.952939	51.74667
dim 9	1.0509244	4.776929	56.52360
dim 10	0.9949498	4.522499	61.04610
dim 11	0.9803760	4.456255	65.50236
dim 12	0.9281096	4.218680	69.72104
dim 13	0.8521577	3.873444	73.59448
dim 14	0.8416311	3.825596	77.42007
dim 15	0.7994716	3.633962	81.05404
dim 16	0.7926129	3.602786	84.65682
dim 17	0.7272327	3.305603	87.96243
dim 18	0.6908054	3.140025	91.10245
dim 19	0.6436571	2.925714	94.02816
dim 20	0.4844556	2.202071	96.23024
dim 21	0.4724870	2.147668	98.37790
dim 22	0.3568612	1.622096	100.00000

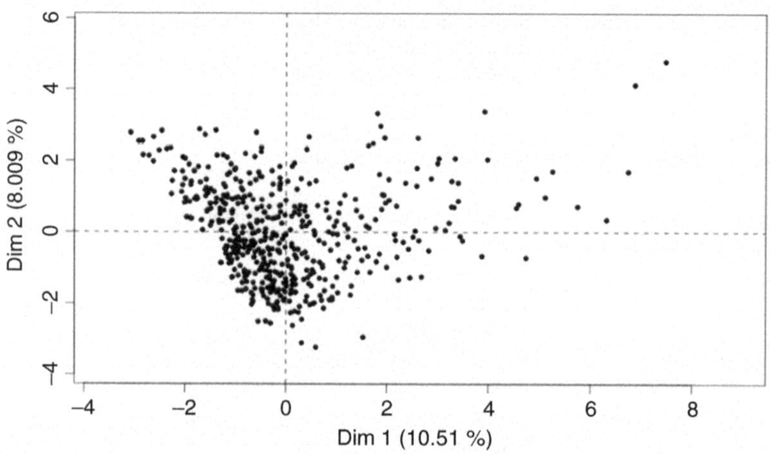

Fig. 14.2 Observations regarding the first two components

the household is highly correlated with component two and the rest of the variables related to household expenditure by item are mostly related to component 1 (Fig. 14.4).

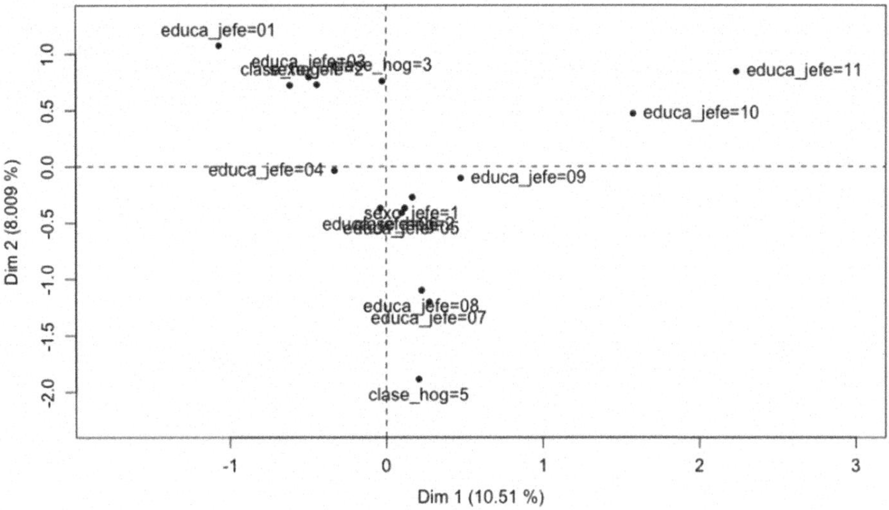

Fig. 14.3 Level map for categorical variables

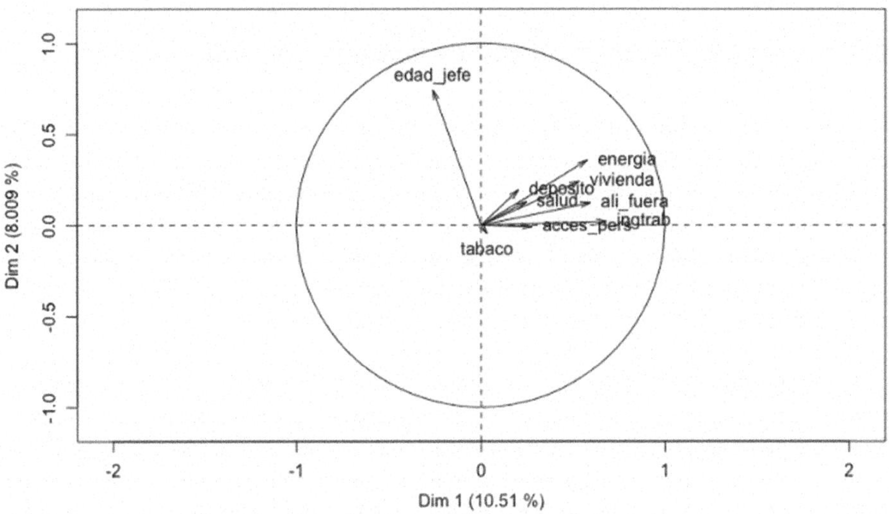

Fig. 14.4 Correlation map of the first two components

The map of the first two main components shows the correlation between the first two main components with all variables (qualitative and quantitative) (Fig. 14.5). For numerical variables, squared charges are squared correlations and for categorical variables square charges are simple correlations.

Properties of the PCA and MCA factor coordinates are applicable in the case of mixed PCA.

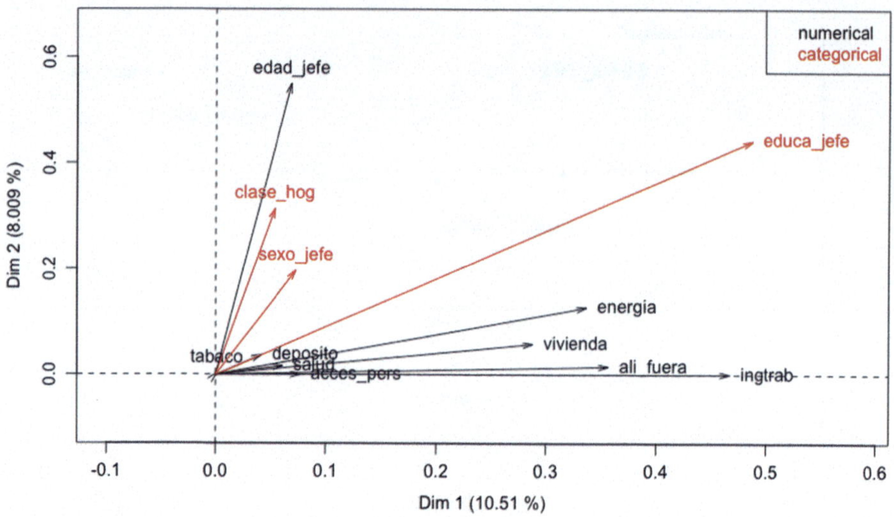

Fig. 14.5 Map of the first two components with all variables

Rotation of the Mixed Principal Component Variables

As in PCA we can use the component variables as axes and perform rotations (the most used is the VARIMAX rotation). The objective of the rotation is to improve the interpretation of the relationship of the main components with both qualitative and quantitative variables, the objective is to maximize the correlation between these variables, graphically it is sought that a set of variables is related to some component clearly [5]. We will use the ENIGH 2016 database with which we have worked. We create a database without the first 100 observations since they will be used later for prediction and validation (Table 14.3).

The results generated have the same interpretation as in the case without rotating, as previously mentioned the objective is to improve the interpretation. The variance explained without change by rotation with three main components (25.11%). The rotation only modifies the squared loads maximizing the correlation between the variables and their components, for a better interpretation. The rotation generates a set of objects that we can use and interpret in a similar way to the results without rotating. To assess the convenience of rotation, we can generate the same result graphs, before and after the rotation. It is possible, exactly as in PCA with mixed data without rotating, to predict the scores of the new observations in the main components after the rotation, and to present the results graphically.

Table 14.3 Eigenvalues from Varimax rotation

Dimension	Eigenvalue	Proportion	Cumulative
dim 1	2.2982381	10.446537	10.44654
dim 2	1.7618783	8.008538	18.45507
dim 3	1.4640838	6.654926	25.11000
dim 4	1.3124019	5.965463	31.07546
dim 5	1.2616679	5.734854	36.81032
dim 6	1.2498359	5.681072	42.49139
dim 7	1.1648184	5.294629	47.78602
dim 8	1.1257336	5.116971	52.90299
dim 9	1.0641733	4.837152	57.74014
dim 10	1.0055071	4.570487	62.31063
dim 11	0.9710124	4.413693	66.72432
dim 12	0.8995455	4.088843	70.81316
dim 13	0.8788191	3.994632	74.80780
dim 14	0.8017024	3.644102	78.45190
dim 15	0.7892203	3.587365	82.03926
dim 16	0.7066036	3.211835	85.25110
dim 17	0.6789509	3.086140	88.33724
dim 18	0.6549098	2.976863	91.31410
dim 19	0.6320754	2.873070	94.18717
dim 20	0.4780495	2.172952	96.36012
dim 21	0.4458733	2.026697	98.38682
dim 22	22 0.3548994	1.613179	100.00000

14.4 Conclusions

Principal component analysis is a multivariate methodology whose main objective is to reduce the size of a set of correlated data in a set of uncorrelated main component variables.

Generally, the methodology corresponding to the sets of variables according to their type is used; if they are qualitative, MCA is used, and if they are quantitative, standard PCA is used. In this work, the PCA methodology with mixed data was presented based on the GSVD methodology where PCA and standard MCA represent a particular case.

The importance of being able to jointly analyze qualitative and quantitative data lies in the representation in a space of smaller dimensions a set of observations, allowing to combine both the numerical and qualitative characteristics of each observation.

The use of the PCAmixdata library implemented in R by Chavent [8] was exemplified, which allows the generation of PCA, MCA, and PCAmix from a data set based on the GSVD methodology.

References

1. J. Amat, Análisis de Componentes Principales (2017), Disponible en https://github.com/JoaquinAmatRodrigo/Estadistica-con-R
2. S. Grossman, *Elementary Linear Algebra* (Cengage Learning, Boston, MA, 1993)
3. F.W. Young, Y. Takane, J. de Leeuw, The principal components of mixed measurement level multivariate data: An alternating least squares method with optimal scaling features. Psychometrika **43**(2), 279–281 (1978)
4. D. Beaton, C.R. Chin Fatt, H. Abdi, An exposition of multivariate analysis with the singular value decomposition in R. Comput. Stat. Data Anal. **72**, 176–189 (2014)
5. M. Chavent, V. Kuentz-Simonet, J. Saracco, Orthogonal rotation in pcamix. ADAC **6**(2), 131–146 (2012)
6. S. Dray, A.-B. Dufour, The ade4 package: Implementing the duality diagram for ecologists. J. Stat. Softw. **22**(4), 1–20 (2007)
7. S. Lê, J. Josse, F. Husson, FactoMineR: An r package for multivariate analysis. J. Stat. Softw. **25**(1), 1–18 (2008)
8. M. Chavent, V. Kuentz-Simonet, A. Labenne, J. Saracco, *Multivariate Analysis of Mixed Data: The R Package PCAmixdata* (Cornell University Library, 2017). Disponible en https://arxiv.org/abs/1411.4911
9. INEGI, Encuesta Nacional de Ingresos y Gastos de los Hogares 2016. ENIGH Nueva Serie. Descripción de la base de datos (2017), Disponible en http://www.beta.inegi.org.mx/contenidos/proyectos/enchogares/regulares/enigh/nc/2016/doc/702825091996.pdf

Chapter 15
SMOTE-Cov: A New Oversampling Method Based on the Covariance Matrix

Ireimis Leguen-deVarona, Julio Madera, Yoan Martínez-López, and José Carlos Hernández-Nieto

Contents

15.1 Introduction 207
15.2 Oversampling Based on the Covariance Matrix 209
15.3 Tools and Experimental Setup 211
15.4 Conclusions and Future Work 213
References 214

15.1 Introduction

Real-world data often present characteristics that affect classification: noise, missing values, inexact or incorrect values, inadequate data size, poor representation in data sampling, etc. The imbalanced dataset problem represents a field of interest as it occurs when the number of instances that represent one class(rare events) [1] is much larger than the other classes, a common problem in certain areas such as fraud detection, cancer gene expressions, natural disasters, software defects, and risk management [2]. Rare events are difficult to detect because of their infrequency and casualness; misclassification of rare events could often result in heavy costs. For example, for smart computer security threat detection [3], dangerous connection attempts may only appear out of hundreds of thousands log records, but failing to identify a serious vulnerability breach would cause enormous losses.

Then, in the case of the datasets with binary class, it can be defined that it is balanced if it has an approximately equal percentage of examples in the concepts to be classified, that is, if the distribution of examples by classes is uniform, otherwise it is unbalanced. To measure the degree of imbalance of a problem, [4] defined the imbalance ratio (IR) as

I. Leguen-deVarona · J. Madera (✉) · Y. Martínez-López · J. C. Hernández-Nieto
University of Camagüey, Camagüey, Cuba
e-mail: ireimis.leguen@reduc.edu.cu; julio.madera@reduc.edu.cu; yoan.martinez@reduc.edu.cu; jose.hernandez@reduc.edu.cu

P. Vasant et al. (eds.), *Data Analysis and Optimization for Engineering and Computing Problems*, EAI/Springer Innovations in Communication and Computing,
https://doi.org/10.1007/978-3-030-48149-0_15

$$IR = \frac{|C+|}{|C-|} \geq 1.5 \quad (15.1)$$

where

$C+$ is the number of instances that belongs to the majority class
$C-$ is the number of instances that belongs to the minority class

Therefore, a dataset is imbalanced when it has a marked difference (IR $\geq$ 1.5) between the examples of the classes. This difference causes low predictive accuracy for the infrequent class as classifiers try to reduce the global error without taking into account the distribution of the data. In imbalanced sets, the original knowledge is usually labeled as oddities or noise, focusing exclusively on global measurements [5]. The problem with the imbalance is not only the disproportion of representatives but also the high overlap between the classes. To overcome this problem, diverse strategies have been developed and can be divided into four groups: at the data level [6, 7], at the learning algorithm level [8], cost-sensitive learning [9], and based on multiple classifiers [10]; where the techniques at the data level are the most used because its use is independent of the classifier that is selected.

One of the best-known algorithms within data-level techniques is the Synthetic Minority Oversampling Technique (SMOTE) [7, 11] for the generation of synthetic instances. One of SMOTE's shortcomings is that it generalizes the minority area without regard to the majority class leading to a problem commonly known as overgeneralization; this has been solved with the use of cleaning methods such as SMOTE—Tomek links (TL) [6, 11], SMOTE—ENN [6, 11], Borderline—SMOTE1 [11, 12], SPIDER [13], SMOTE—RSB* [14], ADASYN [6], among others. These algorithms have been designed to operate with values of both discrete and continuous features for problems with imbalances in their two classes; most of them use the KNN to obtain the synthetic instances, and although this is a method that offers good results, it does not take into account the dependency relationships between attributes, which can influence the correct classification of the examples of the minority class.

A way to obtain the dependency relation of the attributes is Probabilistic Graphical Models (PGMs) [15] that represent joint probability distributions where nodes are random variables and arcs are conditional dependence relationships. Generally, PGMs have four fundamental components: semantics, structure, implementation, and parameters. As part of the PGMs, there are Gaussian Networks that are graphic interaction models for the multivariate normal distribution [16], and some use the Covariance Matrix (CM) to analyze the relationships between variables.

This chapter proposes an algorithm based on SMOTE and the Covariance Matrix estimation to balance datasets with continuous attributes and binary class, exploding the dependency relationships between attributes and obtaining AUC [17] values similar to the algorithms of the state-of-the-art.

An experimental study was performed ranking two SMOTE-Cov variants, SMOTE-CovI (which generates new values within the interval of each attribute) and SMOTE-CovO (which allows some values to be outside the interval of the attributes), against SMOTE, SMOTE-ENN, SMOTE-Tomek Links, Borderline-SMOTE, ADASYN, SMOTE-RSB*, and SPIDER, using 7 datasets from the UCI repository [18] with different imbalance ratios and using C4.5 as a classifier. The performance of the classifier was evaluated using AUC and hypothesis testing techniques as proposed by [19, 20] for statistical analysis of the results.

15.2 Oversampling Based on the Covariance Matrix

This section introduces oversampling based on the Covariance Matrix. First, we describe the Covariance Matrix that allows the computation of variable dependency. Then, we give an overview of our proposed algorithm. Finally, we describe our experimental setup in four steps: tool, dataset selection, evaluation methodology, and classifier used.

Covariance Matrix

In statistics and probability theory, the covariance matrix is a matrix that contains the covariance between the elements of a vector, where it measures the linear relationship between two variables. If the vector-column entries are

$$X = \begin{bmatrix} X_1 \\ \vdots \\ X_n \end{bmatrix} \tag{15.2}$$

then the covariance matrix $\sum_{ij}$ is the matrix, whose (i, j) entry is the covariance

$$\sum_{ij} = \mathbf{E}\left[(X_i - \mu_i)(X_j - \mu_j)\right] \tag{15.3}$$

where the operator **E** denotes the expected value (mean) of its argument

$$\mu_i = \mathbf{E}(X_i) \tag{15.4}$$

The Covariance Matrix allows determining if there is a dependency relationship between the variables and it is also the data necessary to estimate other parameters. In addition, it is the natural generalization to higher dimensions of the concept of the variance of a scalar random variable [20].

SMOTE-Cov

The Algorithm 5 shows the steps of SMOTE-Cov to balance datasets. During the loading of the dataset in the first step, the algorithm expects continuous valued attributes and a binary class. Then, it uses the formula 1 to verify whether the dataset is balanced or not. If it is imbalanced, the algorithm computes the Covariance Matrix. The Covariance Matrix allows the detection of the dependency relationship between attributes. Then, from the estimated covariance matrix, new synthetic instances are generated to balance the minority class. This process stops when an equilibrium between the two classes is reached. The algorithm checks that all the new values generated from the covariance are obligatorily within the interval of each attribute, in the case that some are outside the interval, what is done is to take it to the minimum or maximum, making a kind of REPAIR of the value.

Algorithm 5: SMOTE-Cov steps

```
Input: Dataset X,inRange[Boolean]
Output: Balanced dataset X
Data: Dataset X
Step 1: Load dataset X;
Step 2: Compute X IR using Eq. 15.1;
if IR ≥ 1.5 then
    Step 3: Estimate covariance matrix using Eq. 15.3, this will provide us with a
    probabilistic distribution of the dataset;
    Step 4: For each attribute, a range is determined by it min-max value;
    while X is not in equilibrium do
        Step 5: Generate new instance y according to the covariance matrix;
        if range ≠ true then
            add y to X;
        else
            for i ← 0 to Y_i do
                if Y_i < min Y_i then
                    Y_i = min Y_i;
                else if Y_i > max Y_i then
                    Y_i = max Y_i;
                else
                    continue;
                end
            end
        end
    end
else
    return X;
end
```

15.3 Tools and Experimental Setup

The algorithm was developed using the R language because it is designed for statistical processing and has the cov() function for calculating the covariance. In order to evaluate the behavior of the proposed algorithm, it was compared against the state-of-the-art algorithms of oversampling data balancing; two variants are taken into account: when the attributes are inside or outside of the dependence range. Seven datasets from the UCI repository were chosen with IR ≥ 1.5, see Table 15.1, with continuous attributes and binary class. This experiment uses fivefold cross-validation, and the data are split into two subsets: the training/calibration set (80%) and the test set (20%). The final result is the mean of the 5 result sets. The partitions were made using KEEL in such a way that the number of instances per class remained uniform. The partitioned datasets are available on the KEEL website [21].

The training datasets are balanced, generating new synthetic instances from the minority class to complete the quantities of the majority class and using a sample of the control test, which remains imbalanced and without any modification. The new datasets are generated from the obtained instances, using the SMOTE-Cov algorithm, and a classifier is used as a mean to measure the performance using other techniques.

The classifier used for the experimental study is C4.5 (implemented in the Weka package as J48) [22], which has been referred to as a statistical classifier and one of the top 10 algorithms in Data Mining that is widely used in imbalance problems [14].

The Area Under the Curve (AUC) (15.5) is used to measure the performance of classifiers over imbalanced datasets using the graph of the Receiver Operating Characteristic (ROC) [17]. In these graphics, the trade-off between the benefits (TPrate) and cost(FPrate) can be visualized, which represent the fact that the capacity of any classifier cannot increase the number of true positives without also increasing the false positives. AUC summarizes the performance of the learning algorithm in a number.

$$AUC = \frac{1 + TPrate - FPrate}{2} \tag{15.5}$$

Table 15.1 Description of the datasets used in the experiments

Dataset	Instances	Attributes	IR
ecoli2	336	7	5.4
glass-0-1-2-3_vs_4-5-6	274	9	3.20
glass1	214	9	1.81
Iris	150	4	2
newthyroid2	215	5	5.14
Pima	768	8	1.86
vehicle3	846	18	2.99

where

$TPrate$ are the correctly classified positive cases that belong to the positive class
$FPrate$ are the negative cases that were misclassified as positive examples

Experimental Study

The AUC result values are studied with this already balanced dataset. Table 15.2 shows that the AUC results of the data-balancing algorithm applying the Covariance Matrix with its CovI and CovO variants are similar or comparable with respect to the state-of-the-art oversampling algorithms, using C4.5 as a classifier.

For the statistical analysis of the results, hypothesis-testing techniques were used [19, 20]. In both experiments, the Friedman and Iman-Davenport tests were used [23], in order to detect statistically significant differences between groups of results. The Holms test was also carried out [24] with the aim of finding significantly higher algorithms. These tests are suggested in the studies presented in [19, 20, 23], where it is stated that the use of these tests is highly recommended for the validation of results in the field of automated learning. Table 15.3 shows the ranking obtained by the Friedman test for the experiment. Although the algorithm with the best ranking was ADASYN, Holm's test performed below will demonstrate to what extent this algorithm can be significantly superior to the one proposed in the research.

Table 15.4 summarizes the results of Holms test, taking ADASYN as a control method, all hypotheses with p–values ≤ 0.05 are rejected, showing that ADASYN is significantly superior to the SMOTE-CovI and Borderline-SMOTE algorithms.

Table 15.2 AUC of the data balancing algorithms with the generation of oversampling classes of the state-of-the-art, CovI and CovO

Algorithms	Iris	glass1	Pima	vehicle3	glass-0-1-2-3_vs_4-5-6	ecoli2	new thyroid2
ADASYN	**1**	0.74	0.73	0.74	0.88	0.91	**0.98**
Borderline–SMOTE	0.99	**0.77**	0.70	0.65	0.82	0.89	0.95
SMOTE–ENN	0.99	0.74	0.74	0.71	**0.93**	0.89	0.92
SMOTE–RSB	0.97	0.72	**0.75**	0.73	0.90	0.89	0.96
SMOTE–TL	0.99	0.74	0.72	**0.79**	0.90	0.89	0.93
SMOTE	**1**	**0.77**	0.74	0.72	0.84	**0.92**	0.92
SPIDER	0.99	0.74	0.72	0.71	0.92	0.89	0.95
Original	**1**	0.72	**0.75**	0.72	0.90	0.85	0.96
SMOTE–CovO	**1**	0.71	0.72	0.71	0.92	0.86	0.95
SMOTE–CovI	0.95	0.72	0.70	0.72	0.86	0.86	0.96

The bold values represent the best AUC obtained by each algorithm for each dataset

Table 15.3 Friedman's test

Algorithms	Ranking
ADASYN	3.4286
Borderline–SMOTE	6.9286
SMOTE–ENN	5.4286
SMOTE–RSB	4.9286
SMOTE–TL	5.2857
SMOTE	4.5714
SPIDER	5.6429
Original	5
SMOTE–CovO	6.3571
SMOTE–CovI	7.4286

Table 15.4 Holms test with $\alpha = 0.05$, taking ADASYN as a control method

i	Algorithms	$Z = \frac{(R_o - R_i)}{SE}$	p-value	Holm	Hypothesis
9	SMOTE–CovI	2.47	0.01	0.005	Reject
8	Borderline–SMOTE	2.16	0.03	0.006	Reject
7	SMOTE–CovO	1.80	0.07	0.007	Accept
6	SPIDER	1.36	0.17	0.008	Accept
5	SMOTE–ENN	1.23	0.21	0.01	Accept
4	SMOTE–TL	1.14	0.25	0.012	Accept
3	Original	0.97	0.33	0.01	Accept
2	SMOTE–RSB	0.92	0.35	0.02	Accept
1	SMOTE	0.70	0.48	0.05	Accept

In the case of SMOTE-CovO, SPIDER, SMOTE_ENN, SMOTE_TL, Original, SMOTE-RSB and SMOTE, the null hypothesis is accepted, which means that there are no significant differences between ADASYN and them, so it can be concluded that they are as effective.

15.4 Conclusions and Future Work

In this chapter, a new algorithm is proposed to generate synthetic instances of the minority class, using the Covariance Matrix. The experimental study carried out shows the effectiveness of the proposed algorithm compared to eight recognized state-of-the-art algorithms. SMOTE-Cov showed similar or comparable results, taking into account the results of the AUC curve of the C4.5 classifier and using nonparametric tests to demonstrate that there are no significant differences between them, with the exception of the ADASYN versus the SMOTE-CovI variant. This can be influenced because the attributes present in the studied datasets come from other intervals and not from the actual attribute within the dataset.

Having results comparable to those of the state-of-the-art, these datasets allow extending the experimentation in the future to datasets with tens, hundreds or thousands of attributes and with strong dependency relationships. It is also intended to use covariance regularization (Shrinkage) to balance data, where the number of positive instances is less than the number of attributes.

Acknowledgments We would like to thank VLIR (Vlaamse Inter Universitaire Raad, Flemish Interuniversity Council, Belgium) for supporting this work under the Networks 2019 Phase 2 Cuba ICT and the Play! for food: Improving food production and social welfare in the province of Camagüey project.

References

1. M. Maalouf, T.B. Trafalis, Robust weighted kernel logistic regression in imbalanced and rare events data. Comput. Stat. Data Anal. **55**(1), 168–183 (2011)
2. G. Haixiang, L. Yijing, J. Shang, G. Mingyun, H. Yuanyue, G. Bing, Learning from class-imbalanced data: review of methods and applications. Exp. Syst. Appl. **73**, 220–239 (2017)
3. K. Alzhrani, E.M. Rudd, C.E. Chow, T.E. Boult, Automated big security text pruning and classification, in *2016 IEEE International Conference on Big Data (Big Data)* (IEEE, Piscataway, 2016), pp. 3629–3637
4. A. Fernández, S. García, M.J. del Jesus, F. Herrera, A study of the behaviour of linguistic fuzzy rule based classification systems in the framework of imbalanced data-sets. Fuzzy Sets Syst. **159**(18), 2378–2398 (2008)
5. N.V. Chawla, A. Lazarevic, L.O. Hall, K.W. Bowyer, Smoteboost: improving prediction of the minority class in boosting, in *European Conference on Principles of Data Mining and Knowledge Discovery* (Springer, Berlin, 2003), pp. 107–119
6. G.E.A.P.A. Batista, R.C. Prati, M.C. Monard, A study of the behavior of several methods for balancing machine learning training data. ACM SIGKDD Explorations Newsl. **6**(1), 20–29 (2004)
7. N.V. Chawla, K.W. Bowyer, L.O. Hall, W.P. Kegelmeyer, Smote: synthetic minority over-sampling technique. J. Artif. Intell. Res. **16**, 321–357 (2002)
8. Y.-M. Huang, C.-M. Hung, H.C. Jiau, Evaluation of neural networks and data mining methods on a credit assessment task for class imbalance problem. Nonlinear Anal. Real World Appl. **7**(4), 720–747 (2006)
9. Z.-H. Zhou, X.-Y. Liu, On multi-class cost-sensitive learning. Comput. Intell. **26**(3), 232–257 (2010)
10. M. Galar, A. Fernandez, E. Barrenechea, H. Bustince, F. Herrera, A review on ensembles for the class imbalance problem: bagging-, boosting-, and hybrid-based approaches. IEEE Trans. Syst. Man Cybern. C **42**(4), 463–484 (2011)
11. A. More, Survey of resampling techniques for improving classification performance in unbalanced datasets. (2016, preprint). arXiv:1608.06048
12. H. Han, W.-Y. Wang, B.-H. Mao, Borderline-smote: a new over-sampling method in imbalanced data sets learning, in *International Conference on Intelligent Computing* (Springer, Berlin, 2005), pp. 878–887
13. J. Stefanowski, S. Wilk, Selective pre-processing of imbalanced data for improving classification performance, in *International Conference on Data Warehousing and Knowledge Discovery* (Springer, Berlin, 2008), pp. 283–292
14. E.R. Martínez, F. Herrera, R.B. Pérez, Y.C. Mota, Y.S. López, Edición de conjuntos de entrenamiento no balanceados, haciendo uso de operadores genéticos y la teoría de los conjuntos aproximados

15. D. Koller, N. Friedman, *Probabilistic Graphical Models: Principles and Techniques* (MIT Press, Cambridge, 2009)
16. R.M. Sanner, J.-J.E. Slotine, Gaussian networks for direct adaptive control. IEEE Trans. Neural Netw. **3**(6), 837–863 (1992)
17. A.P. Bradley, The use of the area under the roc curve in the evaluation of machine learning algorithms. Pattern Recognit. **30**(7), 1145–1159 (1997)
18. UCI Machine Learning Repository, https://archive.ics.uci.edu/ml/index.php. Accessed 14 Feb 2019
19. S. García, F. Herrera, Evolutionary undersampling for classification with imbalanced datasets: proposals and taxonomy. Evol. Comput. **17**(3), 275–306 (2009)
20. D.J. Sheskin, *Handbook of Parametric and Nonparametric Statistical Procedures* (CRC Press, Boca Raton, 2003)
21. KEEL-dataset repository, http://www.keel.es/. Accessed 14 Feb 2019
22. J.R. Quinlan, *C4. 5: Programs for Machine Learning* (Elsevier, Amsterdam, 2014)
23. R.L. Iman, J.M. Davenport, Approximations of the critical region of the fbietkan statistic. Commun. Stat. Theory Methods **9**(6), 571–595 (1980)
24. S. Holm, A simple sequentially rejective multiple test procedure. Scand. J. Stat. pp. 65–70 (1979)

Chapter 16
A Sentiment Analysis Method for Analyzing Users Opinions About Drugs for Chronic Diseases

María del Pilar Salas-Zárate, Giner Alor-Hernández, Jorge Luis García-Alcaraz, Luis Omar Colombo-Mendoza, Mario Andrés Paredes-Valverde, and José Luis Sánchez-Cervantes

Contents

16.1 Introduction 217
16.2 Related Work 219
16.3 Sentiment Analysis Method for Drug Opinions Analysis 220
16.4 Evaluation 222
16.5 Conclusions and Future Work 226
References 227

16.1 Introduction

The information on the Web is growing dramatically, which is leading to a series of challenges that range from security and privacy issues to technical and technological issues. At the same time, it is turning the Web into a vast source of information

M. del Pilar Salas-Zárate (✉) · G. Alor-Hernández · L. O. Colombo-Mendoza
M. A. Paredes-Valverde
Tecnológico Nacional de México/I. T. Orizaba, Orizaba, Veracruz, México
e-mail: msalasz@ito-depi.edu.mx; galor@ito-depi.edu.mx; lcolombom@ito-depi.edu.mx; mparedesv@ito-depi.edu.mx

J. L. García-Alcaraz
Department of Industrial Engineering, Universidad Autónoma de Ciudad Juárez, Ciudad Juárez, Chihuahua, Mexico
e-mail: jorge.garcia@uacj.mx

J. L. Sánchez-Cervantes
CONACYT-Tecnológico Nacional de México/I. T. Orizaba, Orizaba, Veracruz, México
e-mail: jlsanchez@conacyt.mx

P. Vasant et al. (eds.), *Data Analysis and Optimization for Engineering and Computing Problems*, EAI/Springer Innovations in Communication and Computing,
https://doi.org/10.1007/978-3-030-48149-0_16

that can be exploited to derive actionable knowledge about virtually anything. In this context, as the years go by, more and more people use the Web, e.g., blogs, forums, wikis, social networks, and review websites, as the primary medium for answering their queries and making decisions in multiple domains, not being the exception the healthcare context. Considering that chronic diseases are prevalent medical conditions in low and middle,[1] it makes sense to think that most of the Web queries about medical conditions are related to chronic diseases. In fact, according to the National Center for Chronic Disease Prevention and Health Promotion of the USA, 6 in 10 adults in that country have one chronic disease.[2] Similarly, according to the National Institute of Geography, Statistics, and Informatics of Mexico, by 2015, diabetes was the leading cause of death among Mexican adults above 60 years.[3] Hence the importance of developing computational methods and tools to automatically derive actionable knowledge about chronic diseases that is valuable not only for patients but also for doctors, health professionals, and health authorities from Web resources.

Sentiment analysis is an on-going and fast-growing subfield of computer science that analyzes users' opinions about different topics such as individuals, organizations, products, services, events, as well as their attributes [1]. It has been successfully applied to analyze user-generated content on the Web in several different domains. Sentiment analysis can benefit the healthcare domain for developing multiple methods and tools such as methods for detecting adverse drug reactions mentions [2, 3], opinion summarization and extraction systems for drugs and doctors [4, 5], and recommender systems for health self-management [6, 7].

Several authors have introduced methods to analyze and process opinions from social media data. Most of these efforts are based on machine learning (ML) and semantic orientation (SO) approaches. The ML approach requires a set of data to train an algorithm and built a predictive model; and a set of data for evaluate the built model. On the other hand, the SO approach is based on lexicons (SentiWordNet, ML-Senticon, WordNet-Affect, etc.). However, there are several linguistic features expressed in opinions that they are not considered; for example, features related to cultural, social, psychologic aspects can have an important impact on sentiment analysis.

This work proposes a sentiment analysis method, which combines machine learning with psycholinguistic features to analyze opinions of users about drugs and determine their polarity. For this purpose, this method uses the LIWC tool [8], which already has been used to analyze the emotional well-being of people by analyzing their social media posts [9, 10] and to identify language features that distinguish demographic and psychological attributes from language in blogs [11]. An experiment was specifically conducted over a corpus of Mexican patient

[1] https://www.who.int/nutrition/topics/2_background/en/.

[2] https://www.cdc.gov/chronicdisease/resources/infographic/chronic-diseases.htm.

[3] https://www.inegi.org.mx/contenidos/saladeprensa/aproposito/2017/muertos2017_Nal.pdf.

opinions about drugs for diabetes and hypertension that we compiled through an ad-hoc drug information website.

This work is structured as follows: Sect. 16.2 presents the relevant literature on sentiment analysis in the health self-care domain. Then, Sect. 16.3 describes the sentiment analysis method, whereas Sect. 16.4 describes a case study that shows its application for analyzing opinions about drugs. Finally, we present our conclusions and future directions in Sect. 16.5.

16.2 Related Work

Diverse sentiment analysis approaches in the form of methods, tools, and platforms, ranging from machine learning-based approaches to semantic orientation-based approaches, have been reported in the literature recently. Machine learning-based approaches mostly employ supervised machine learning algorithms; these algorithms initially fit a model from the features of the documents in a corpus using a training dataset and require a second dataset called test dataset to evaluate a final model built from the training dataset. In regard to semantic orientation-based approaches, sentiment lexicons such as SentiWordNet, iSOL, and eSOL [12] are employed. A basic sentiment analysis task is sentiment polarity detection, often called sentiment polarity classification, which involves identifying the polarity of an opinion expressed at the level of a document, sentence, entity, or aspect using a scale consisting of three degrees: positive, negative, or neutral [13].

With the aim of identifying the opportunity niches of our research, we analyzed and contrasted some works on sentiment analysis, specifically, on sentiment polarity detection, in the domain of chronic diseases and associated drugs.

Jiménez-Zafra et al. [14] studied how people express their opinion about drugs and doctors in medical forums in Spanish. They used both a machine learning approach and a semantic orientation-based approach to polarity detection over two different corpora extracted from the forums mimedicamento.es and masquemedicos.com for that purpose. In particular, the support vector machines algorithm and the iSOL sentiment lexicon were, respectively, used as the foundation of these approaches. A probabilistic aspect mining approach for drug reviews is presented in [15]. This approach differs from most approaches to aspect level-based opinion mining in the sense that it is not aimed at extracting the aspects and their sentiments from opinions but at identifying the aspects related to class labels or categorical meta-information in opinions. In fact, it is more related to topic modeling than to aspect level-based opinion mining. Gopalakrishnan et al. [16] studied patient satisfaction with drugs using online drug reviews. A machine learning-based approach to polarity detection that relies on artificial neural networks was proposed for that purpose. According to the results of an experiment performed on a corpus of drugs opinions obtained from the askapatient.com website, this approach performs better than other related approaches when the Radial Basis Function Neural Network model is specifically used. A study of the impact of sentiment analysis features in detecting adverse drug reaction (ADR) mentions in tweets and forum posts was presented in [3]. In this study, a semantic orientation-based approach to polarity detection was integrated into an algorithm for ADR mentions extraction, namely a supervised conditional random field classifier for sequence labeling that is called ADRMine. A corpus of tweets and posts from support groups or forums of the DailyStrength social network was used by the authors for validation. Biyani et al. [17] analyzed the sentiment of users posts about cancer in English. In particular, the co-training machine learning algorithm was used to conduct polarity detection as a classification task based on both domain-dependent sentiment features and domain-independent sentiment features. A corpus of posts from the Cancer Survivors' Social Network (CSN) of the American Cancer Society was used by the authors for validation. Similarly, in [18] a sentiment analysis method and a tool to detect the polarity of the posts written by cancer patients in Brazilian online cancer communities were presented. This method corresponds to a semantic orientation-based approach and uses the dictionary of Portuguese terms used by the SentiStrength sentiment analysis tool. A series of experiments were conducted by the authors for validation purposes using a corpus of posts collected from the Facebook social network. The authors also proposed in a previous work [12] an aspect-level method for sentiment analysis of tweets about diabetes written in English. In particular, this method performs polarity detection at the aspect level by (1) semantically annotating tweets to identify aspects using an ontology, namely the diabetes diagnosis ontology and (2) detecting the polarity of each of the identified aspects using the SentiWordNet lexicon.

Based on the related work discussion above presented, next conclusions can be derived: (1) there is a lack of works on sentiment analysis methods for analyzing patient opinions about chronic disease medicines written in Spanish; (2) most of the works proposed in this domain that use hybrid approaches to polarity detection do not study the effect of psychological features but only the effect of linguistic features; and (3) no evaluation or validation on real-world datasets of Mexican patient opinions about drugs for diabetes and hypertension is reported in the literature on sentiment analysis. We intend to address these issues as will be seen later in this paper.

16.3 Sentiment Analysis Method for Drug Opinions Analysis

The sentiment analysis method proposed in this work relies on machine learning and psycholinguistic analysis technologies to analyze the user's opinions on drugs. As depicted in Fig. 16.1, before any analysis can proceed, a corpus of opinions is necessary. Then, corpus processing, psycholinguistic features extraction, and learning process are performed. Next sections thoroughly describe these phases.

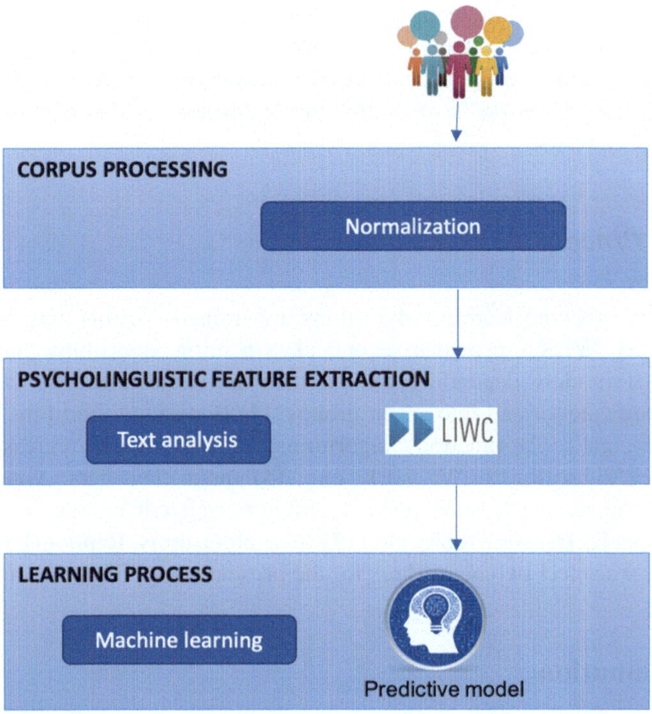

Fig. 16.1 Sentiment analysis method

Corpus Processing

Corpus processing. In the health self-care domain, most opinions about drugs are written by non-expert people. Therefore, the language used in these opinions is characterized by informal writing with spelling errors. Aiming to make the corpus collected more useful for linguistic research purposes, it is subjected to a process where spelling errors are corrected. Hunspell tool is used for finding spelling errors within an opinion. In this way, words such as "daibetes" were corrected to "diabetes" (diabetes). Hunspell was selected because it has been successfully applied for text preprocessing purposes in the sentiment analysis domain [19–21].

Psycholinguistic Feature Extraction

In this paper, we used the LIWC tool to obtain features such as psychologic linguistic of the text. LIWC carries out an analysis based on five main macro-categories (linguistic, psychological, personal concerns, spoken, and punctuation marks).

LIWC has been used in sentiment analysis only as a lexicon based on semantic orientation approach, which is constructed from two subcategories (positive emotions and negative emotions) of the 72 subcategories of LIWC [22] and [23]. However, only a few works have studied the importance of this type of features in the sentiment analysis.

Learning Process

In this phase, machine learning algorithms are trained. To this aim, WEKA tool [24] was used. WEKA provides several classification algorithms that allow the generation of models depending on the data and objective. The classification algorithms are categorized into seven groups: (1) Bayesian algorithms: BayesNet, Naive Bayes, etc.; (2) functions algorithms: logistic, linear regression, SMO, etc.; (3) lazy algorithms: LWL, IBk, etc.; (4) meta classifiers: Vote, Bagging, etc.; (5) miscellaneous: InputMappedClassifier, SerializedClassifier, etc.; (6) rules algorithms OneR, DecisionTable, etc.; (7) trees algorithms: RandomTree, J48, etc. Specifically, we select two algorithms for the present study, BayesNet and SMO.

16.4 Evaluation

In this section, the evaluation process is described. The objective is to measure the effectiveness of the proposed sentiment analysis method regarding users' opinions classification in the context of drugs for diabetes and hypertension. The below sections describe the people involved and corpus used in this work and discuss the obtained results.

Subjects

For the purposes of this case study, 20 people with diabetes and/or hypertension were asked to interact with the Web application aiming to collect opinions about drugs they use for managing the aforementioned diseases (see Fig. 16.2). All participants were native Spanish speakers, ranged from 35 to 60 years old. Furthermore, these people have lived with diabetes and/or hypertension for many years which make them suitable for this work because of their experience with the use of drugs for managing these chronic diseases.

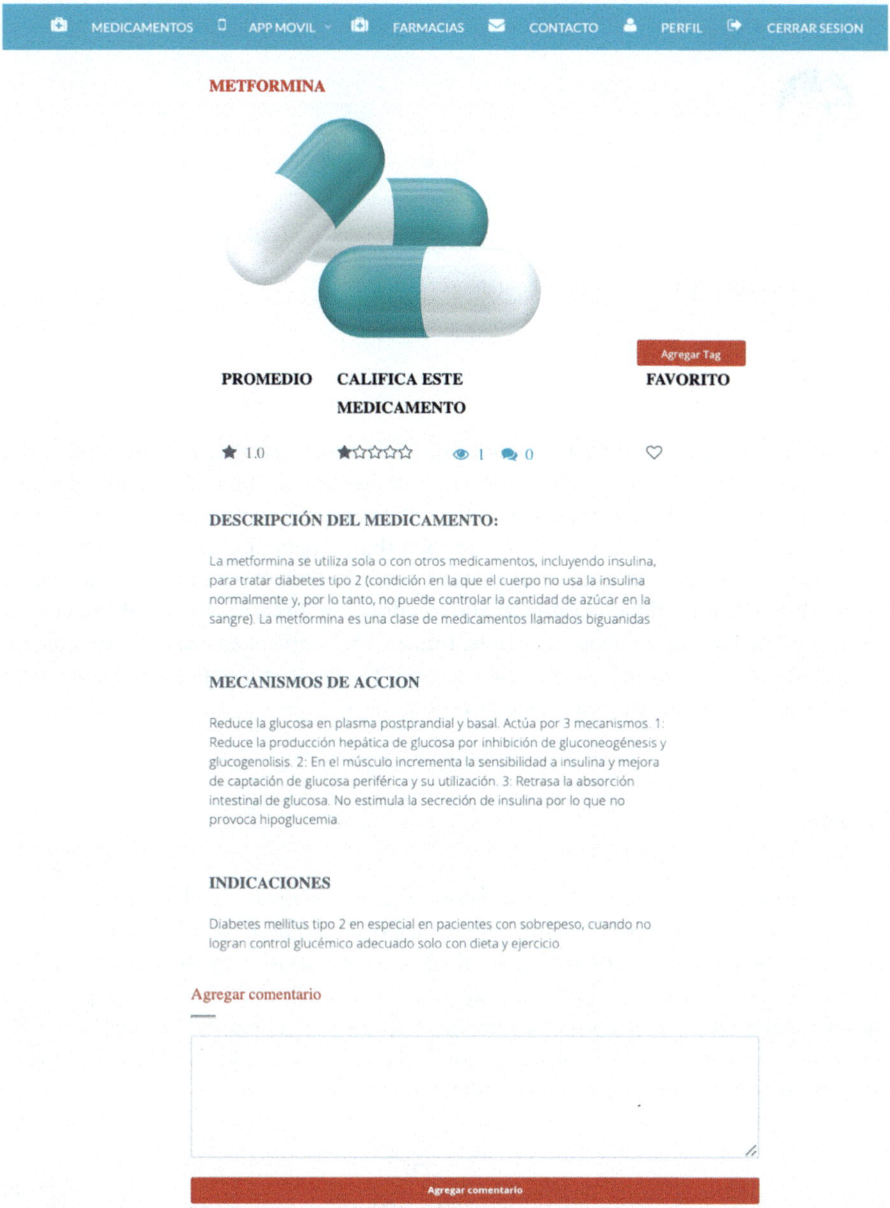

Fig. 16.2 Web interface

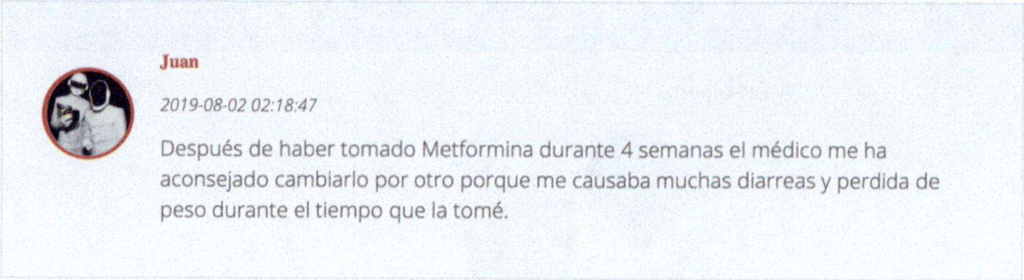

Fig. 16.3 Example of user's opinion about drugs

Corpus

The case study described in this work was performed using a corpus of reviews written in Spanish that describe users view or judgment about drugs for diabetes and hypertension. This corpus was collected through a web platform of chronic diseases (see Fig. 16.2). It must be mentioned that a manual review of all collected opinions was carried out aiming to discard reviews that seem out of the scope of this work or do not reflect an opinion about a drug. In this way, the resulting corpus consists of 740 opinions about 24 drugs. Figure 16.3 depicts an example of a user's opinion about "metformin" drug, which is used for the treatment of Diabetes type 2 in adults that helps control blood sugar levels.

Results

In the present work, recall (R), precision (P), and F-measure (F1) metrics were selected to measure the performance of the proposed method. Recall (see Eq. (16.1)) represents the number of correctly classified opinions divided by the number of data that should be identified as positive, negative, or neutral; Precision (see Eq. (16.2)) represents the number of correctly classified opinions divided by the number of all opinions returned by the classifier as correctly classified; and the F-measure (see Eq. (16.3)) is the harmonic mean of precision and recall.

$$R = \frac{\mathrm{TP}}{\mathrm{TP} + \mathrm{FN}} \tag{16.1}$$

$$P = \frac{\mathrm{TP}}{\mathrm{TP} + \mathrm{FP}} \tag{16.2}$$

$$F1 = 2 * \frac{\mathrm{Precision} * \mathrm{Recall}}{\mathrm{Precision} + \mathrm{Recall}} \tag{16.3}$$

where

Table 16.1 Evaluation results with SMO algorithm

	2 classes			3 classes		
LIWC categories	*P*	*R*	*F*1	*P*	*R*	*F*1
Linguistic process	0.713	0.712	0.712	0.657	0.657	0.657
Psychological process	0.736	0.736	0.736	0.649	0.768	0.704
Personal concerns	0.581	0.566	0.573	0.556	0.527	0.541
Spoken categories	0.528	0.513	0.520	0.514	0.514	0.514
Punctuation marks	0.633	0.595	0.613	0.546	0.604	0.574
Combination of all categories	0.794	0.793	0.793	0.754	0.81	0.781

Table 16.2 Evaluation results with BayesNet algorithm

	2 classes			3 classes		
LIWC categories	*P*	*R*	*F*1	*P*	*R*	*F*1
Linguistic process	0.698	0.698	0.698	0.653	0.715	0.683
Psychological process	0.708	0.708	0.708	0.675	0.738	0.705
Personal concerns	0.593	0.591	0.592	0.552	0.552	0.552
Spoken categories	0.553	0.553	0.553	0.52	0.52	0.520
Punctuation marks	0.588	0.569	0.578	0.502	0.578	0.537
Combination of all categories	0.726	0.725	0.725	0.697	0.738	0.717

True Positive (TP) are those positive, negative, or neutral opinions that the system classified as such; False Negatives (FN) are those positive, negative, or neutral opinions that the classifier not detected as such; and False Positives (FP) are the opinions that the system classified into another different class than the one it belongs to.

Tables 16.1 and 16.2 present the results obtained with SMO and BayesNet algorithms for the sentiment analysis based on psycholinguistic features.

As can be observed, the general results show that the best features for classification of drug opinions are psychological and linguistic features. On the other hand, the features that do not contribute to better classification are personal concerns and spoken features. We ascribe this to two reasons: (1) opinions contain a great number of grammatical words that are part of the linguistic category; (2) opinions contain words related to the feelings of the users. It is important to mention that people with a chronic illness such as diabetes usually suffer from a mental illness such as anxiety or depression, which can be reflected in their opinions.

On the other hand, the best results are obtained with the combination of all LIWC categories with an F-measure of 79.3%.

The results also show that SMO algorithm is more accurate for this classification problem than BayesNet. This result can be due to the study presented in [25], where authors conclude that SMO is more robust and accurate than other algorithms. Finally, the classification with 2 classes (positive opinions and negative opinions) provides better results than the classification with three classes (positive opinions, negative opinions, and neutral opinions). Therefore, the results demonstrate that with fewer categories the classification algorithm performs better.

16.5 Conclusions and Future Work

In the health self-care domain, most opinions about drugs are written on the Web by consumers themselves, i.e., by people who are not necessarily medical experts. Therefore, informal media such as social networks and user review websites become prominent sources of consumer opinions about drugs. Presumably, one of the hottest topics in this context are drugs for chronic diseases.

Sentiment analysis or opinion mining can be exploited to automatically derive actionable knowledge about drugs for chronic diseases from consumer opinions on the Web that is valuable not only for the patients themselves but also for doctors, health professionals, and health authorities.

Chronic diseases are among the most prevalent medical conditions around the world, and the Americas is not the exception. Nonetheless, there is a lack of works on sentiment analysis methods for analyzing patient opinions about chronic disease medicines written in Spanish. In this piece of research, we intended to address this issue by proposing a hybrid sentiment analysis method that combines machine learning features with psycholinguistic features to determine the polarity of patient opinions about drugs written in Spanish.

We validated our method using a corpus of Mexican patient opinions about drugs for diabetes and hypertension that was compiled through a drug information website that we developed for the purposes of this study. We obtained hopeful results with a F1-measure score of 0.793, which support our assumption that using a combination of linguistic and psychological features yields better results than using linguistic features solely. In fact, the combination of all the five categories of LIWC, namely linguistic, psychological, and personal concerns, spoken categories, and punctuation marks yielded better results relation to F1-measure values in comparison with each one of the individual categories.

Furthermore, regarding the number of polarity classes employed in our experiment, the use of two classes, namely positive and negative, yielded better results. This can be interpreted as evidence that opinions about chronic diseases drugs on the Web are highly polarized among Mexican patients.

As regards future work, we have planned to integrate new techniques to improve the sentiment analysis method. To this end, we will analyze the possibility to use an ontology that represents semantic knowledge about chronic diseases and related drugs to identify aspects in patient opinions and conduct sentiment analysis at the aspect level. In this context, calculating the polarity of consumer opinions about drugs at the sentence level does not necessarily results in identifying what a consumer likes or dislikes about a drug, because the same sentence can actually contain multiple opinions about multiple features of one single drug, e.g., cost and side effects. We will also study the use of topic models, e.g., Latent Dirichlet Allocation, to conduct the aspect mining process as a topic (or rather, sub-topic) discovery process.

Likewise, considering that we can find consumer opinions about drugs written in Spanish on multiple sources on the Web, we plan to integrate new data sources, specifically, social networks, e.g., Twitter and Facebook, to validate our proposal using diverse opinion corpora.

Finally, the proposed sentiment analysis method may be integrated into complex applications or systems that can be especially valuable in the healthcare domain, specifically, for health self-management purposes; recommender systems and question answering systems are great examples of these complex systems.

Acknowledgments This work was supported in part by the National Council of Science and Technology of Mexico (CONACYT) and the Secretariat of Public Education of Mexico (SEP) through the Professional Development Program for University Professors (PRODEP). Additionally, Tecnológico Nacional de Mexico (TecNM) supported this research.

References

1. B. Liu, Sentiment analysis and opinion mining. Synth. Lect. Hum. Lang. Technol. **5**, 1–167 (2012)
2. A. Sarker, G. Gonzalez, Portable automatic text classification for adverse drug reaction detection via multi-corpus training. J. Biomed. Inform. **53**, 196–207 (2015)
3. I. Korkontzelos, A. Nikfarjam, M. Shardlow, A. Sarker, S. Ananiadou, G.H. Gonzalez, Analysis of the effect of sentiment analysis on extracting adverse drug reactions from tweets and forum posts. J. Biomed. Inform. **62**, 148–158 (2016)
4. R. Rivas, N. Montazeri, N.X. Le, V. Hristidis, Automatic classification of online doctor reviews: Evaluation of text classifier algorithms. J. Med. Internet Res. **20**, e11141 (2018)
5. D. Cavalcanti, R. Prudêncio, Aspect-Based Opinion Mining in Drug Reviews, in ed. by E. Oliveira, J. Gama, Z. Vale, H. Lopes Cardoso, *Progress in Artificial Intelligence, EPIA 2017, Lecture Notes in Computer Science*, vol 10423 (Springer, Cham, 2017)
6. R.L. Rosa, G.M. Schwartz, W.V. Ruggiero, D.Z. Rodriguez, A knowledge-based recommendation system that includes sentiment analysis and deep learning. IEEE Trans. Ind. Inf. **15**, 2124–2135 (2019)
7. D. Yang, C. Huang, M. Wang, A social recommender system by combining social network and sentiment similarity: A case study of healthcare. J. Inf. Sci. **43**, 635–648 (2017)
8. J.W. Pennebaker, M.E. Francis, R.J. Booth, Linguistic inquiry and word count: LIWC 2001. Mahw. Lawrence Erlbaum Assoc. **71**, 2001 (2001)
9. L. Zhang, X. Huang, T. Liu, Z. Chen, T. Zhu, Using Linguistic Features to Estimate Suicide Probability of Chinese Microblog Users, in ed. by Q. Zu, B. Hu, N. Gu, S. Seng, *Human Centered Computing, HCC 2014, Lecture Notes in Computer Science*, vol 8944 (Springer, Cham, 2014)
10. M. Settanni, D. Marengo, Sharing feelings online: studying emotional well-being via automated text analysis of Facebook posts. Front. Psychol. **6**, 1045 (2015)
11. T. Yarkoni, Personality in 100,000 words: A large-scale analysis of personality and word use among bloggers. J. Res. Pers. **44**, 363–373 (2010)
12. M. Del Pilar Salas-Zárate, J. Medina-Moreira, K. Lagos-Ortiz, H. Luna-Aveiga, M.Á. Rodríguez-García, R. Valencia-García, Sentiment analysis on tweets about diabetes: An aspect-level approach. Comput. Math. Methods Med **2017**, 1–9 (2017)
13. E. Cambria, B. Schuller, Y. Xia, C. Havasi, New avenues in opinion mining and sentiment analysis. IEEE Intell. Syst. **28**, 15–21 (2013)
14. S.M. Jiménez-Zafra, M.T. Martín-Valdivia, M.D. Molina-González, L.A. Ureña-López, How do we talk about doctors and drugs? Sentiment analysis in forums expressing opinions for medical domain. Artif. Intell. Med. **93**, 50–57 (2019)

15. V.C. Cheng, C.H.C. Leung, J. Liu, A. Milani, Probabilistic aspect mining model for drug reviews. IEEE Trans. Knowl. Data Eng. **26**, 2002–2013 (2014)
16. V. Gopalakrishnan, C. Ramaswamy, Patient opinion mining to analyze drugs satisfaction using supervised learning. J. Appl. Res. Technol. **15**, 311–319 (2017)
17. P. Biyani, C. Caragea, P. Mitra, C. Zhou, J. Yen, G.E. Greer, K. Portier, Co-training over domain-independent and domain-dependent features for sentiment analysis of an online cancer support community, in *Proceedings of the 2013 IEEE/ACM International Conference on Advances in Social Networks Analysis and Mining - ASONAM'13* (ACM Press, New York, 2013), pp. 413–417
18. R.G. Rodrigues, R.M. Das Dores, C.G. Camilo-Junior, T.C. Rosa, SentiHealth-cancer: A sentiment analysis tool to help detecting mood of patients in online social networks. Int. J. Med. Inform. **85**, 80–95 (2016)
19. J.M. Cotelo, F.L. Cruz, J.A. Troyano, F.J. Ortega, A modular approach for lexical normalization applied to Spanish tweets. Expert Syst. Appl. **42**, 4743–4754 (2015)
20. F.H. Khan, S. Bashir, U. Qamar, TOM: Twitter opinion mining framework using hybrid classification scheme. Decis. Support Syst. **57**, 245–257 (2014)
21. M.T. Wiley, C. Jin, V. Hristidis, K.M. Esterling, Pharmaceutical drugs chatter on Online Social Networks. J. Biomed. Inform. **49**, 245–254 (2014)
22. P. Gonçalves, M. Araújo, F. Benevenuto, M. Cha, Comparing and combining sentiment analysis methods, in *Proceedings of the First ACM Conference on Online Social Networks - COSN'13* (ACM Press, New York, 2013), pp. 27–38
23. C.J. Hutto, E. Gilbert, VADER: A parsimonious rule-based model for sentiment analysis of social media text, in *Eighth International AAAI Conference on Weblogs and Social Media* (Georgia Institute of Technology, Atlanta, GA, 2014)
24. R.R. Bouckaert, E. Frank, M.A. Hall, G. Holmes, B. Pfahringer, P. Reutemann, I.H. Witten, *WEKA—Experiences with a Java Open-Source Project* (MIT Press, Cambridge, 2001)
25. H. Bhavsar, A. Ganatra, A comparative study of training algorithms for supervised machine learning. Int. J. Soft Comput. Eng. **2**, 2231–2307 (2012)

Chapter 17
Proofs of the Undecidability of Stegananalysis Techniques

Juan Gutierrez-Cardenas

Contents

17.1 Introduction 229
17.2 Background: Turing Machines and Steganography 230
17.3 On the Undecidability of Steganalysis Techniques: Proofs 233
17.4 Conclusions 240
References 241

17.1 Introduction

Steganography and Steganalysis are two techniques complementary to each other, in which the first one aims to hide information on diverse types of media, while the second one attempts to find this concealed information. During the years, we have foreseen the development of different steganographic and steganalysis techniques. Nevertheless, steganalysis techniques are far from being perfect, because they have always suffered from a rate of false positives when trying to detect hidden information, or payload, inside a media [9]. For this reason, the development of more accurate steganalysis techniques has become an active field of research in the last years. Equally important, it is essential to mention the relevance of security scenarios in which steganalysis procedures have been used, from research studies in the field of security and terrorism [12], to the business field along with several software solutions trends in the past few years [6].

In the following research work, we will show that the derivation of a general or universal steganalysis tool is an undecidable problem. For proving this, we will employ the techniques of Cantor Diagonalization as well as Reduction theory. Furthermore, we based our research on the articles of Chess and White [2] and Cohen [3], where the authors established the impossibility of finding a program that

J. Gutierrez-Cardenas (✉)
Universidad de Lima, Lima, Peru
e-mail: jmgutier@ulima.edu.pe

P. Vasant et al. (eds.), *Data Analysis and Optimization for Engineering and Computing Problems*, EAI/Springer Innovations in Communication and Computing,
https://doi.org/10.1007/978-3-030-48149-0_17

could detect a set of computer viruses. Based on this premise, we will try to prove the impossibility of the existence of a general steganalysis tool that will be capable of detecting a payload deceived into a cover object.

We organized this chapter in the following parts: In Sect. 17.2, we present the concepts of steganography and steganalysis techniques along with a brief definition of Turing machines, in Sect. 17.3 we aim to establish the hypothesis that states that the generation of an innocuous computer virus has some simile characteristics with a steganography procedure. Additionally, we will prove that steganalysis procedures are undecidable problems by using the Diagonalization of Cantor and the reduction technique, finishing with the conclusions of the present research.

17.2 Background: Turing Machines and Steganography

Steganography Definition

Steganography is a technique that strives to conceal information by masking it inside an object that acts as cover media. For this sort of camouflage purpose exists a plethora of different methods—starting with modification of the bit's information of the cover media to the application of algebraic or signal transformations. In addition, the utilization of unused parts that form part of the cover object [15], e.g., least significant bits on a picture or flag bits on IP packets, has become a common practice in steganography. An example of this procedure is depicted in Fig. 17.1.

According to Khalind, Hernandez-Castro, and Aziz [9], there exist two types of steganographic procedures: spatial domain and transform domain. The Spatial domain manipulates the pixels that form part of the cover image, while the Transform domain performs alterations on the signal or into the statistical properties of the cover image. In the latter case, the usual methods applied are the Discrete Cosine or the Discrete Wavelet transform [14]. In general terms, a space transformation steganography such as the Least-significant-bit or LSB will perform the following steps:

Fig. 17.1 General schemata of a steganographic procedure: a cover image will act as a carrier for the data to conceal. The data to hide, in this case, the cat figure, will set its pixel information on the cover image by modifying the bit information of their pixels. Another steganographic procedure could perform image transformations on the cover image. In either case, the result is a stego-object that contains the concealed data, which is not visible by simple inspection

```
Algorithm 1
1. bit_representation ← transform(data_to_hide)
2. while length(bit_representation)>0
3.    map(hide(bit_representation[length −1]),cover_image)
4.    length(bit_representation) minus 1
5. end while
```

While a Domain transformation technique, as the one that employs Haar transforms by using wavelet transformations, could follow these steps [10]:

```
Algorithm 2
1. Employ the 2D Haar Transform
2. Find a set of coefficients under a certain threshold
3. Replace the bits of step 2 with the bits of the
   message to hide
4. Apply the inverse transform
5. Store the transformed data
```

As we observe, Algorithms 1 and 2 perform a certain degree of bit manipulation to the cover images, by using a direct transformation of the bits or by relying on the transformation of the statistical properties of the images. In either case, there exists a sort of object manipulation, and based on that, we can state the following theorem:

Theorem 17.1 *Every steganography transformation oriented to images, which has a space or domain transform, will perform a modification in the pixel information of the cover image.*

This theorem would prove to be useful when we would aim to establish the undecidability of steganalysis techniques in a general way, but before reaching that part we need to explain the general concept of steganalysis.

Steganalysis is the counterpart of steganography; similarly, there are plenty of steganalysis techniques as there exist steganographic ones. The classification of the former ones is [8]:

1. Stego-only, in which only the stego-object is available.
2. Known-cover or known-message attack.
3. Chosen-stego or chosen-message attack.
4. Blind-steganalysis

Being the latter one is the hardest form for detecting a payload. The reason is because in that schema, the attacker does not have information about the stego-object, the payload, or the cover object. So, there exists only an assumption that an object could hide a payload on them.

Definition of a Turing Machine

A Turing Machine (TM) is a mathematical model of a computer that will allow, under certain conditions, to execute an algorithm on it. The conditions must be that a procedure should be evaluated in a finite form, not to mention under discrete steps, as defined in Hopcroft and Ullman [7], see Fig. 17.2.

Hopcroft and Ullman defined a Turing machine as a computational model in which a header reads, writes, or modifies the information that is on a tape. The header can access the information from left to right and vice versa. Furthermore, each cell could contain a finite number of symbols. In section "Steganography and Turing Machines", we will derive a general steganographic procedure, based on a Turing Machine.

Undecidability

According to Poonen [11] there exist two types of problems that can be considered undecidable:

1. Axiom independent: We have this case when it is impossible to deduct the definition of what is evaluated or its negation by using a set of logic rules and axioms.
2. Decision problems. In this case, the examined problem should return a Yes/No answer, but the difficulty here is that if we evaluate an algorithm, this will not halt and return a correct answer each time. For our proving purposes, we will consider this category of problems.

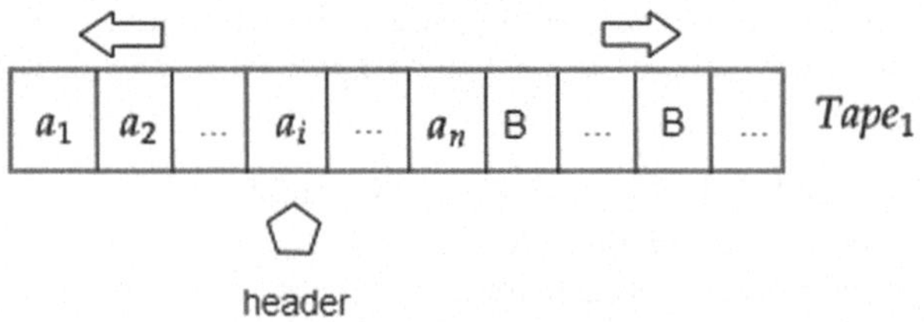

Fig. 17.2 Schemata of a Turing machine (adapted from Hopcroft and Ullman [7]). In this schema, we have a tape that contains a set of symbols a_i and blank spaces B. A header, represented by c, traverses the tape reading or writing symbols on the tape by moving it in the right or left direction

17.3 On the Undecidability of Steganalysis Techniques: Proofs

For this proof, we will demonstrate the impossibility of the existence of a universal steganalysis tool that will be capable of detecting a payload hidden into a cover image. For our purposes, first, we will model a general steganographic algorithm by a TM. Second, we will compare our model and the basics of a steganography model to the actions that are conducted by an innocuous computer virus. Third, we will use the diagonalization of Cantor and undecidability via reduction techniques, from the definition of a computer virus, to prove that the problem of finding a universal steganalysis algorithm is undecidable.

Steganography and Turing Machines

Steganography algorithms can be depicted by using a TM, as we can see in Fig. 17.3. The premise here is that in any type of steganographic procedure, spatial or in domain transform performs changes to the internal structure of a cover image.

For exemplification purposes, we will be focusing on the spatial domain transform, arguing that our proposal will not lose generality in the case of a spatial or domain transform. We hypothesize that in either case, we have altered the information that is on the cover image. This modification of the cover image allows us to store a hidden message or payload, and for revealing it, a receiver should extract this information from the modified pixels. We should also consider that the present model can be extended to other media, such as the case of network steganography, with the only difference that IP packets replace the cover image.

In general terms, our schema will be formed by the tuple: Displayed equations are centered and set on a separate line.

$$StegoObject = S(CoverObject, Mensaje, [Key]) \tag{17.1}$$

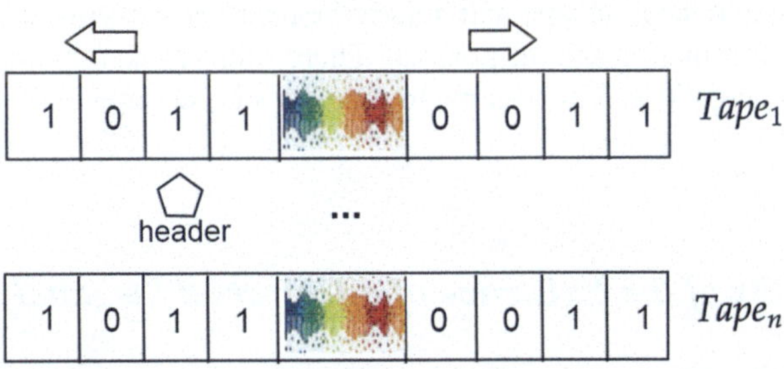

Fig. 17.3 Turing machine representing a steganography algorithm. On a tape we have the pixel information of the cover object. By moving the header in a left or right direction, we can modify the bit information, which could be concealed by a spatial domain transform, or attached to the image transforms by using Wavelets or Cosine transforms. In either case, manipulation of pixels or block of pixels is given, together with transformations into the cover image

In this tuple, the result is a steganography object or stego object that stores a message within a cover image or cover object. In our schema, a sender can use an encryption key to map the data to hide with the pixel information of the cover object. Nevertheless, it is valuable to mention that the use of a key is optional. According to the schema shown in Fig. 17.3, we can have a tape or set of tapes that contain information about the original values of the pixels of the cover image, while another tape can store the posterior modification of the cover image values. This obfuscation part is performed by considering that a header is overwriting the information on the tape or storing the modified information in other tapes. The modification of the image information of the cover image will allow it to hold a payload. The inverse procedure of recovering the original message can also be made with another Turing Machine. This TM would reverse the steganography procedure leading to the extraction of the hidden information, which was encoded in the form of bits on a tape. Consequently, we can model this generalized procedure of information hiding by using a representation that consists of making changes to the bit information, which is present on a tape to hold a payload. For example:

Let's consider q_0 as the initial state; this node will point to any arbitrary bit position recorded on the tape. Then we can establish the following values for starting a steganographic procedure:

```
Initial tape information:       1101
Payload:                        1110
Objective tape information:     0011
```

Then, we can have the following set of moves to change the information on the tape:

$q_0 10 R q_0$
$q_0 10 R q_0$
$q_0 01 R q_0$
$q_0 11 R q_0$

In this sequence, we suppose the information or payload will be hidden in the first four nonsignificant bits of a pixel. For illustration purposes, we have considered that the information that the final tape will hold is the result of applying an XOR between the initial tape information and the payload. Therefore, we can hypothesize, by using this proof of concept, that it is possible to model a simple steganographic algorithm by using a TM.

Undecidability of the Existence of a Universal Steganalysis Tool

In this part of our research, we will give the preliminary algorithms that will help us to show the impossibility of the existence of a general steganalysis tool. First, we will derive an algorithm that describes the general steps of any steganographic procedure:

```
Algorithm 3: Generalization of a steganography procedure
S(Cover Object, Message, Key)
{Key is an object that contains the probable positions to
modify within the cover object at a pixel's level}
1. for i←1 to Message.length
2.   CoverObject←modifyCover(Cover_Object,Key(Message[i]))
3. endfor
4. return CoverObject
```

Algorithm 3 picks up each part of the message we want to hide, together with applying a key for encrypting the message or for giving a mapped location within the cover image. In the end, the cover image is modified in its pixel information. This algorithm can be transformed into a form of innocuous viral form as it was defined by Cohen [4]. Cohen established that not all viral forms are dangerous. For example, a program that uncompresses a file when a bit reaches some value. We have devised a sketch of this harmless viral form as follows:

```
Algorithm 4: Innocuous computer Virus
1. Loop over an uninfected file
2.        if virus_id=true
3.                 new_file←compress file ∪ file_C
4.        else
5.                 do_nothing
6. end loop
```

As we can observe Algorithm 4 finds an uninfected file, compresses it, and adds another file into it. In this case, the resulting file contains a sort of payload, but without delivering any harm to the original file. Our modification, based on Algorithm 4, would be the following; resembling a steganography procedure:

```
Algorithm 5: Modification of a generic steganography
algorithm so that it is similar to an innocuous computer
virus.
obtainStegoObject
2. loop coverImage.getPixel(i)
2.        key_sub_value←Key.getPixelValue(i)
3.        if getPixel(i)=key_sub_value
4.                 extract x_i from the message, apply change,
                   coverObject.setPixel(x_i)
5.        endif
6. end loop
```

Algorithm 5 modifies the value of each pixel trying to adapt a payload on it, having a simile to a noninfectious computer virus. The position of each pixel to be transformed could be determined by the key value while looping over each of the selected pixels from the cover image. One could claim we can build a direct mapping considering the positions of the pixels, but we consider the basics of a steganographic procedure, as we showed in Algorithm 3.

Another point to acknowledge is that steganography is not only used for illicit purposes, but it can also be used for making marks on digital media, which will allow knowing whom the original author of this material was. This procedure is known as watermarking and furthermore could be considered as an innocuous use of steganography in a similar fashion of what Cohen made with his model of virus for data compression.

According to a survey made by Cheddar et al. [1], we can deduce that a steganography algorithm will always perform some bit manipulation inside the cover image. Considering this, we could formally design a Turing machine that hides information into the pixels of a cover image. Let's remember that a cover image, as any digital image, is formed by pixels that can be transformed into binary data according to the information of the RGB channels. Our model will be the following:

$$M = \left\{Q, \Sigma, \Gamma, \delta, q_0, q_{\text{accept}}, q_{\text{reject}}\right\} \tag{17.2}$$

where q_0 is the initial state so the header will point at the beginning in this position. Also in $\{0, 1, \Sigma, \xi\}$ a bit will be represented by 0 or 1, and the symbol ξ will mark an empty space on the tape.

Our alphabet will serve when we concatenate the data to represent the information of a pixel on a cover object. The pixel contents are affected by some manipulation in an attempt to conceal any hidden information on it. We believe that a TM for representing a steganography or steganalysis procedure fails in the category of a Deterministic TM. This assumption stands as a result of both procedures working as an injective function. If this will not be the case, then a steganalysis procedure will give more than one value in the uncovering of a message, thus leading to wrong information recovered.

In the following algorithm, we also show how to perform a simple LSB for hiding a payload. For this case, we are considering two tapes, one holding the pixel information of the cover image, and the other that contains the information, in bits, we want to hide inside the cover image:

Algorithm 6: TM for the LSB steganography technique
M1 is a TM that will serve to conceal a message
Precondition: $w \in input \wedge cover\_image \subseteq \Gamma \wedge payload \subseteq \Gamma^*$
1. $headerPayload \in \Gamma \leftarrow lefmost\ position \neq empty$
2. $headerPayload \in \Gamma^* \leftarrow lefmost\ position$
3. while header or headerPayload point to a position≠null
4.   write on Γ the data corresponding to what is pointing the headerPayload
5.   move the header on the type four positions to the right
6.   move the headerPayload one position to the right
7. endwhile
8. if headerPayload=null then stops

Also, we could use a finite automaton to describe the sequence of states, which follow the aforementioned description. Starting from an initial state, this FA will overwrite the data on the tape, considering the information of the payload. In each overwrite operation, this automaton will traverse to a next state, verifying if the tape of the payload is empty, so, in this case, it will reach an ending state.

In the following section, we will try to demonstrate that to find a tool that will allow us the detection if an image contains a payload to be undecidable.

Proofs

We will construct two proof sketches to prove that the problem of finding a general steganalysis tool is undecidable: the first one applies the principle of diagonalization of Cantor, and the second one applies reduction. But before that, we will draft a proof of undecidability by using the concept of an oracle in a TM.

Theorem 17.2 *The problem to find a universal steganalysis tool is undecidable.*

Proof Let us assume that S_{TM} is a set of steganalysis algorithms supported by a Turing Machine given that a certain input M was applied to a cover object C_{obj} for hiding a message. So, we have that:

$$S_{\mathrm{TM}} = \{< M, < C_{obj} >> s.t\ M\ is\ a\ \mathrm{TM}\ and\ M\ detects\ it\ if\ C_{obj}\ contains\ a\ payload\}$$

Now we define a Turing Machine H, which acts as a decider of S_{TM}, such that:

$$H < M, < C_{obj} >>= \begin{cases} \text{accept,} & \text{if } M\ detects\ a\ payload\ in\ C_{obj} \\ \text{reject,} & \text{if } M\ does\ not\ detect\ a\ payload\ in\ C_{obj} \end{cases}$$

In the same way, we could define an additional Turing machine D_{TM}, this TM will contain H as a subprogram. In addition, this D when given some input returns an answer opposed to the results of H, see Fig. 17.4.

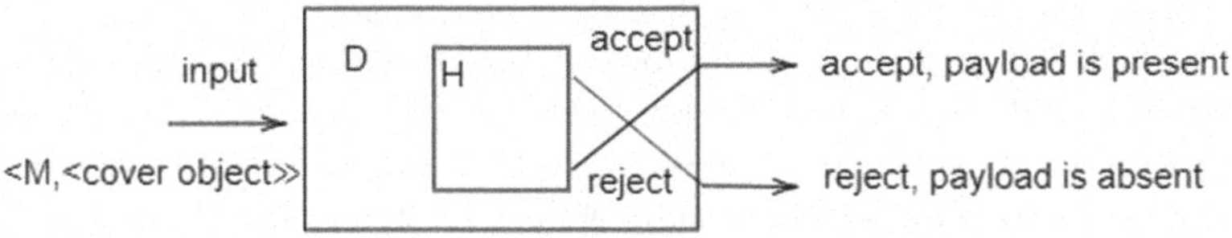

Fig. 17.4 A Turing Machine D, which contains a subroutine of H, the output obtained from H is the opposite of D

According to Fig. 17.4, when we execute H with the input S_{TM}, we will have the following results:

$$D < M, < C_{obj} >>= \begin{cases} \text{accept,} & \text{if } C \textit{ does not contains a payload} \\ \text{reject,} & \text{if } C \textit{ contains a payload} \end{cases}$$

In this last instance, we could replace the input of D with $< M >$ and make D to execute it with the input $< D >$, which corresponds to the same program. This last assumption is not incompatible with the concept of a Turing machine, considering that a TM will be able to run any other algorithm so it can also execute another TM. For our case, we will assume the rules established in the description of Fig. 17.4 still holds. This will give us:

$$D << D >>= \begin{cases} \text{accept,} & \text{if } D \textit{ does not contains a payload} \\ \text{reject,} & \text{if } D \textit{ contains a payload} \end{cases}$$

In this scenario, we have a contradiction because we noticed that D_{TM} does not contain a conclusive answer. Instead, it is a contradictory one that fails to determine if the input contains a payload or not. According to that, we can say that our S_{TM} is undecidable.

Proof By Diagonalization of Cantor:

Let us design a schema as the one explained by Sipser [13], which is also applied for the undecidability of virus detection; as mentioned in the research work of Cohen [3, 4] and Chess and White [2]. Let us make a matrix in which the rows M_i are several steganalysis algorithms supported by n Turing Machines. The labels of the columns will be bit chains that represent the pixel information of a cover image that could contain or not a payload. In this case, an acceptance situation will occur when the Turing machine detects effectively that a chain contains hidden information; while a reject situation will occur when the Turing machine determines that the input does not contain a payload. When we examine the table, we can observe that could exist a machine $M_k < M_n$; where M_n represents an infinite countable set of Turing machines that run a steganalysis algorithm. M_k represents a TM that when receiving as input each one of the chains $< M_n >$, detects the opposite of the results we have found in the diagonal of the obtained matrix. As we can observe, this machine M_k has not been defined by any of the inputs, represented by the distinct TM given until this moment; therefore, we can state that M_k is inexistent (Fig. 17.5).

For example, let us evaluate the following case: If we have that for machine M_1 we have an input that consists of the machine $< M_2 >$ we will have an acceptance output (a), but for the inverse case of the machine M_2 input $< M_1 >$ we will have a reject output (r). An undecidable situation will occur in the case of machine D with input $< D >$ because any input will lead to a contradiction of the expected behavior that we will expect from that machine. Explained in other terms, this rejection occurs as a result that there is no TM such that $M_i = M_k$ and this can occur in the following cases:

	$< M_1 >$	$< M_2 >$	...	D
$< M_1 >$	a	r	...	a
$< M_2 >$	a	r	...	a
...	...			
D	r	r	...	?
...	...			

Fig. 17.5 Diagonalization technique. In this case, we have a set of Turing Machines M_i that accepts different inputs that correspond to the same Turing machine or its description $< M_i >$. We can have a TM D, which by having $< M >$ as an input will output the opposite value as M when submitted to the same input. The contradiction will occur when D takes as an input its own description $< D >$

(a) If $< M_i > \in M_k \; then < M_i > \notin M_i$
(b) If $< M_i > \notin M_k \; then < M_i > \in M_i$

Situations that, in any case, direct us to a contradiction. So, in conclusion, the TM M_k will be inexistent; moreover, we can argue that the set of TM $M(w)$ will be non-recursively enumerable. This kind of proof has its principles in the diagonalization theorem of Cantor.

Proof By Reduction:

We can attempt to make a reduction from the Undecidability of Detection of Computer Viruses stated by Cohen [3, 4]. As we know, the undecidability of virus detection is a problem that can be solved by reduction to the Halting problem. In this case, starting from the algorithm of a general or universal steganalysis tool, we will show that the algorithm devised would help to build an instance of a type of the virus detection problem. Therefore, we should start with the general theorem regarding the problem of virus detection:

Theorem 17.3 *The problem of virus detection is undecidable.*

Proof This theorem has been proven by Cohen [4].

Theorem 17.4 *The problem of payload detection by steganalysis techniques is undecidable.*

Proof Let us start with a program P that was obtained by using Algorithm 5 mentioned before, which, as we recall, was outlined from the modification of an algorithm that produces an innocuous virus. The premise was that a program $MAKE\_INNOCUOUS\_VIRUS()$ would perform changes to a file, but will not generate a state of $INFECT()$ in the system [5]. Algorithm 5 performs a similar task; it modifies bits to conceal a message, but with no harm or infection of the system at all. According to Algorithm 5, P would modify an input w, which is a cover object, so that it would conceal a hidden message. Moreover, we can have a program $STEGO\_DETECT(P)$ that would return a value of true if it detects a payload in w obtained by P. Formalizing this would be:

$$HALT(P) = (STEGO\_DETECT(P(w) \rightarrow concealed\_message(w)))$$

which is similar to the definition that arises when we have a program that tries to detect a computer virus and fails to do so.

Additionally, we can have a program P that concealed a message, w, on a cover image, but that $STEGO\_DETECT$ fails to find a payload. In this case, the program will not halt; therefore, it will loop indefinitely returning a value of true. This value will make the function HALT(P) to return an incorrect answer. Therefore, the steganalysis tool examined is undecidable.

The reduction can also be derived from the undecidability of the Halting problem. For example, let's consider the following pseudo-code, which halts if we have a function that detects a hidden payload on a cover object:

```
halts(p)
        if is_forged((detect_payload(image))) return true
        else return false
```

$is\_forged(p)$ is a program that returns true if the image has been modified, so $detect\_payload(image)$ will also return true as a result of finding a hidden message inside the cover image. Furthermore, we can have a function $detect\_payload(input)$ that fails to detect if an image contains a payload, and we can call this function $failed\_detect\_payload(image)$. This function will loop indefinitely until a payload is detected, so in this case, the definition of halts(p) would be as follows:

```
halts(p)
      if is_forged((not(detect_payload(image))) return true
      else return false
```

As we can observe our program will fail to return a correct answer that would serve to halt the program, because of that our algorithm for detecting a payload will result in an undecidable problem.

17.4 Conclusions

We have demonstrated the undecidability of the developing of a general or universal steganalysis tool. The proofs we have shown in this research are based on the principle of Cantor diagonalization, reduction techniques to the halting problem, and the undecidability of computer viruses. We were able to show that there is a simile between the steganographic process of hidden information and the schemata of an innocuous computer virus. This is an effort to show the impossibility of devising a general steganalysis tool. The undecidability of steganalysis tools does not mean that the efforts and research devised toward the detection of hidden information in different media would be a fruitless one. Nevertheless, we should aim further efforts to increase the research in this field, given that using these techniques, one of the uttermost valuable entities in our modern world, information, could be compromised.

References

1. A. Cheddad, J. Condell, K. Curran, P.M. Kevitt, Digital image steganography: survey and analysis of current methods. Signal Process. **90**, 727–752 (2010). https://doi.org/10.1016/j.sigpro.2009.08.010
2. D.M. Chess, S.R. White, An undetectable computer virus, in *Proceedings of Virus Bulletin Conference*, Orlando, 2000
3. F. Cohen, Computer viruses: theory and experiment. Comput. Secur. **6**, 22–35 (1987). https://doi.org/10.1016/0167-4048(87)90122-2
4. F. Cohen, Computational aspects of computer viruses. Comput. Secur. **8**, 325–344 (1989). https://doi.org/10.1016/0167-4048(89)90089-8
5. D. Evans, On the impossibility of virus detection. University of Virginia (2017), https://www.cs.virginia.edu/~evans/virus/. Accessed 26 Oct 2018
6. J. Gutiérrez-Cárdenas, Steganography and data loss prevention: an overlooked risk? Int. J. Secur. Its Appl. **11**(4), 71–84 (2017). https://doi.org/10.14257/ijsia.2017.11.4.06
7. J.E. Hopcroft, J.D. Ullman, R. Motwani, *Introduction to Automata Theory, Languages and Computation*, 3rd edn. (Pearson Education, Harlow, 2006)
8. S. Katzenbeisser, F. Petitcolas, *Information Hiding Techniques for Steganography and Digital Watermarking* (Artech House Books Inc., Norwood, 1999)
9. O. Khalind, J. Hernandez-Castro, B. Aziz, A study on the false positive rate of Stegdetect. Digit. Investig. **9**(3), 235–245 (2013)
10. H. Lo, S. Topiwala, J. Wang, Wavelet based steganography and watermarking (2016), http://www.cs.cornell.edu/topiwala/wavelets/report.html. Accessed 26 Oct 2016
11. B. Poonen, Undecidable problems: a sampler, in *Interpreting Gödel: Critical Essays*, Chap. 10, ed. by J. Kennedy (Cambridge University Press, Cambridge, 2014), pp. 211–241. http://www-math.mit.edu/~poonen/papers/sampler.pdf
12. N. Provos, P. Honeyman, Detecting steganographic content on the internet, in *Proceedings of Network and Distributed System Security Symposium (NDSS)*, 2002
13. M. Sipser, *Introduction to the Theory of Computation*, 3rd edn. (Cengage Learning, Boston, 2013)
14. H. Shivaram, D. Acharya, R. Adige, P. Kamath, A secure color image steganography in transform domain. Int. J. Cryptogr. Inf. Secur. (2013). https://doi.org/10.5121/ijcis.2013.3103
15. M. Warkentin, E. Bekkering, M.B. Schmidt, Steganography: forensic, security, and legal issues. J. Digit. Forensics Secur. Law **3**(2), 17–34 (2008)

Chapter 18
Vulnerability of Network Coding Under Pollution Attacks

Raúl Antonio Ortega Vallejo and Francisco de Asís López Fuentes

Contents

18.1 Introduction 243
18.2 Related Work 245
18.3 Methodology 246
18.4 Implementation and Evaluation 246
18.5 Conclusions 252
References 252

18.1 Introduction

Today many interactive applications such as videogames or video streaming generate a great traffic in Internet. This large amount of traffic has created problems on the communications networks, such as delays and a bad quality in the received files. In this context, network coding can do important contributions to the information transmission schemes for computer networks, because this technique allows optimizing the transfer of packets between the nodes [1]. The procedure of this scheme consists in using the potential of the intermediate nodes in the network, which interact to encode the information coming from the source nodes. In this way, the receiving nodes can decode and recover the original message, competing with traditional forwarding schemes [2, 3]. It is shown in Fig. 18.1, where source S transmits x to node N1, and y to node N2. Both nodes retransmit the same flow to N3, which using the operator xor ($\otimes$), sends the flow to N4, where it is retransmitted to the receivers R1 and R2. For example, when the packet is received in R1, it does the operation $x \otimes (x \otimes y)$ to obtain y. R2 does a similar operation to obtain x. Here, network coding allows to save time in the receiving nodes, without exceeding the limits of the capacity for each channel, which transmit one bit per unit of time [2,

R. A. O. Vallejo · F. de Asís López Fuentes (✉)
Department of Information Technology, Universidad Autónoma Metropolitana – Cuajimalpa, Mexico City, Mexico
e-mail: flopez@correo.cua.uam.mx

P. Vasant et al. (eds.), *Data Analysis and Optimization for Engineering and Computing Problems*, EAI/Springer Innovations in Communication and Computing,
https://doi.org/10.1007/978-3-030-48149-0_18

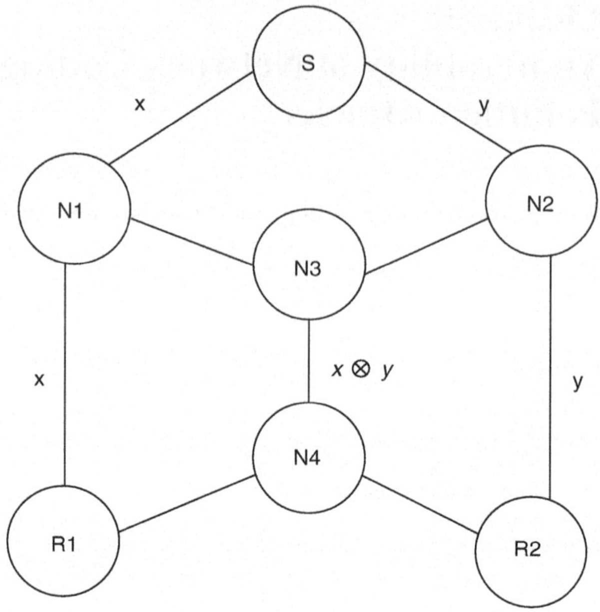

Fig. 18.1 Network coding with butterfly topology [1]

3]. It is also important to mention that network coding has been mainly oriented to increase throughput and reliability of communication networks [4].

The constant contribution to the issue of network coding has led to the emergence of paradigms that propose an improvement to connectivity schemes. This should also formulate security approaches that involve using network coding in communication systems. However, network coding scheme without security methods implies the fact that the flow of information travels susceptible to espionage, and alteration of it. In other words, attackers can modify the packets and send modified packets without any limitation. Network coding is susceptible to attacks that can influence in the loss of data, and cause damage to the integrity of the scheme. In this paper, we review the pollution attack [5] which generates inefficient communication between the nodes. This type of attack consists in the injection of corrupt packets during the processes of reception, coding, and decoding of the authentic packets, in such a way that the corrupt packets are taken as valid packets. This means that the attacker first visualizes the flow of the network, choosing the edge (that would be the connection channel), and intercepts a packet. The packet is modified and injected, taking the place of the legitimate packets, and causing the nodes to continue coding with the corrupt packets [5].

In this paper, we evaluate the pollution attack effects [6] in network coding using traditional scheme called butterfly. We want to emphasize the importance of the absence of security mechanisms in network coding schemes. In other words, network coding looks a great opportunity into the research projects about information theory and network optimization, but, at the same time, has produced certain questions. These points must be studied in order to reduce the probability of occurrence of attacks on network coding.

The rest of this paper has been organized as follows. In Sect. 18.2 we give a brief description about related works. In Sect. 18.3 we explain our methodology used in our experiments to evaluate the pollution attack in a network, using network coding. Evaluation of a prototype and the obtained results are described in Sect. 18.4. Our conclusions and some ideas as future work are presented in Sect. 18.5.

18.2 Related Work

Research on the use of network coding has allowed generating more contributions, increasing academic interest to explore concepts that can strengthen latent points or discussion of the subject, in this case the implications in the security context. The network coding is susceptible to attacks that can influence the loss of data and cause damage to its integrity. For example, a pollution attack [5] can generate inefficient communication between nodes, in addition to causing loss of information.

The pollution attack is one of the most frequent attacks for data coding [5]. This type of attack consists of the injection of corrupt packets, that is, in the processes of receiving, encoding, and decoding the authentic packets from the source, the corrupted packets are taken as valid packets. This means that the attacker first visualizes the network flow, choosing the edge (connection channel). In consequence the interception of the packet (here the importance of the information content enters if the adversary uses Gaussian eliminations, he can access it). The packet is modified and injected, taking the place of the corresponding legitimate packet, and causing the nodes to continue coding with the corrupt packet.

A security scheme for network coding is KEPTE [7], which is a scheme focused on a pre-distribution of keys, based on the coding of tags (coming from the coding of the data), using algebraic methods. Currently, the scheme has been criticized [8] for vulnerability issues presented in its original publication [7]. The main idea of KEPTE is to have two vectors (one public and one private). Thus, there is a process for a transmission generation, in this case becoming a KDC (key distribution center), in order to distribute two secret keys to each node, where said vectors are stored. After that, a tag generation process is performed, and then the verification of packets with n tags is performed. The problem of KEPTE is that the model only focuses on a single key generation transmission; this makes the scheme vulnerable to modern attacks, such as mass attacks. Guangjun [8] proposes a strengthening of the model, using a dynamic generation of secret keys.

The authors in [9] propose SPOC (safe practical network coding). This model focuses on the use of methods that contemplate specific cryptographic algorithms (for example, the AES algorithm—advanced encryption standard) in order to hide information from an adversary, under RLNC (random linear network coding). The main idea of this proposal is to handle the concept of the existence of two coefficient vectors, one in "locked" mode and another in "unlocked" mode, naturally the locked vector contains the information to be protected. These vectors are processed by the intermediate nodes, with the advantage that by using strong tools such as AES (where the size of the encrypted text is equivalent to that of the plain text), they allow discarding the existence of effects on the protocol used in the scheme of the network. In this way, the message can be decrypted on the receiving nodes.

18.3 Methodology

To evaluate the impact of pollution attack in a communication scheme with network coding, we have used the following methodology:

(a) Implement a network scheme with network coding to carry out a security analysis. Therefore, the network environment will be implemented, with this we refer to the code programming necessary for the operation of the scheme. Consequently, it is ideal to think of a small network. For example, under our design, contemplate $2n$ connected devices, where $(1 \leq n \leq 8)$. With the purpose of having clearer and more understandable scenarios for the analysis, as well as for its justification. Our network coding implementation is based on the butterfly scheme.
(b) Once the computer network with network coding is implemented, different types of pollution attacks will be applied in different points of our network coding scheme. That is, in the same way, said process contemplates the programming of code that is required.
(c) Different experiments will be carried out for different scenarios, where the necessary conditions exist for an attack of contamination to occur. The objective is interpret the profile of an adversary, following the inject process of corrupt packets.
(d) Analyze the results obtained and validating all observations may help us to define the real effect of a pollution attack at different points in the network.

18.4 Implementation and Evaluation

A network coding prototype based on butterfly scheme has been implemented in our lab. Our implementation is carried out in the Operating System Debian 4.9, using the C programming language. Also, we implement threads for each multicast process to simulate a double communication at the same time (for example, for the intermediate nodes). In our experiments, we create a virtual scheme framework using Linux containers to simulate the infrastructure of a butterfly network. Each container works like a node (socket connection). First, we need to implement the traditional network coding model, and then create small programs for simulate the attacks. The case 1 (Fig. 18.2) illustrates the typical scheme of network coding. Both receivers can decode all messages from the sources N1 and N2, means the sinks R1 and R2 can recover all packets without problem. In this case, nodes N1

Fig. 18.2 Case 1

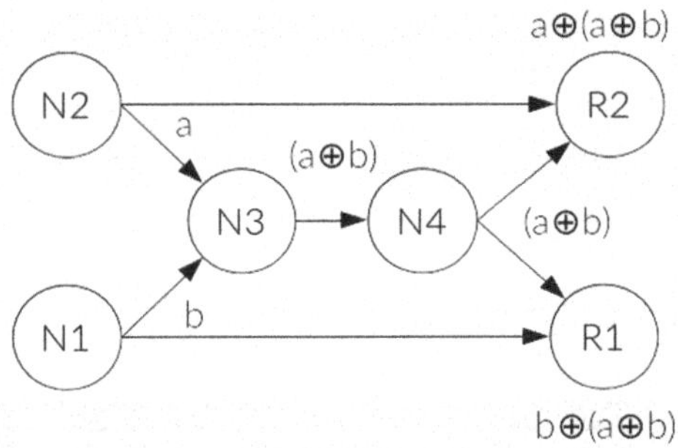

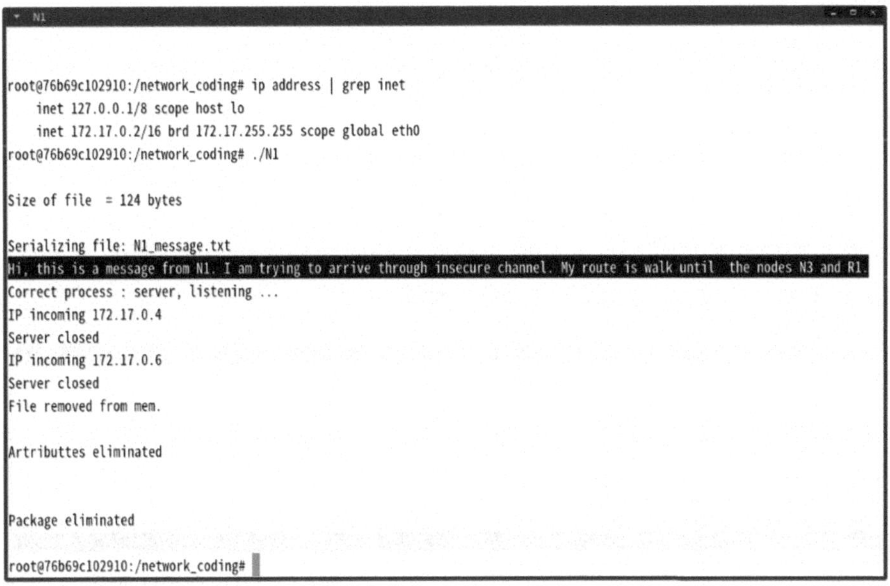

Fig. 18.3 Message sent from Node N1 to nodes R1 and R2

and N2 operate as servers. R2 connects with N2, while R1 connects with N1. Node N3 connects with N1 and N2, then mix the packets A and B (a XOR operation is used). After that, N3 operates as server, sending the compute $A \otimes B$ inside a packet with destiny to N4, which forwards information to nodes R1 and R2. Finally, R1 and R2 decode to retrieve messages from both nodes, N1 and N2.

To evaluate this scenario in our implementation, the node N1 sends this message "Hi, this is a message from N1. I am trying to arrive through insecure channel. My route is walk until the nodes N3 and R1." to node R1 and R2 (see Fig. 18.3). On the other hand, node N2 sends this message "Information is the resolution of uncertainty. Claude Shannon." to node R1 and R2 (see Fig. 18.4).

Node N3 uses network coding to combine both messages as $A \otimes B$ and transmits this message to node N4, which distributes message $A \otimes B$ to nodes R1 and R2. Then, R1 does operation $A \otimes (A \otimes B)$ to recover message from node N2, while

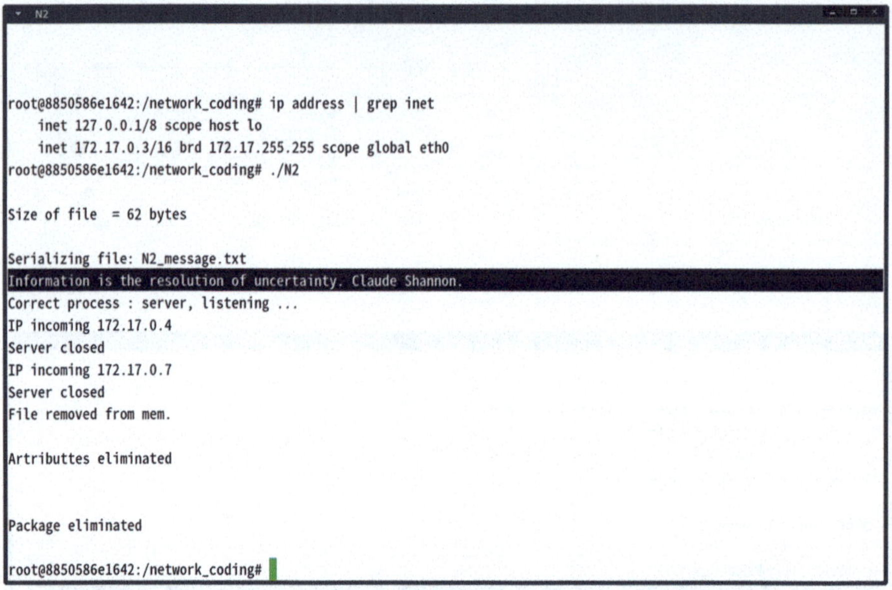

Fig. 18.4 Message sent from node N2 to nodes R1 and R2

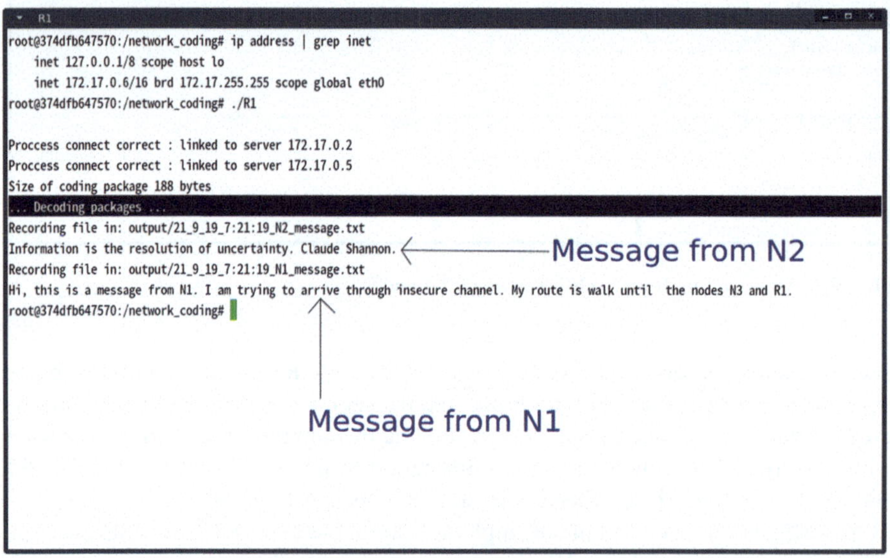

Fig. 18.5 Message recovered in node R1

node R2 does operation $B \otimes (A \otimes B)$ to recover message from node N1. Messages recovered in node R1 and node R2 are shown in Figs. 18.5 and 18.6, respectively.

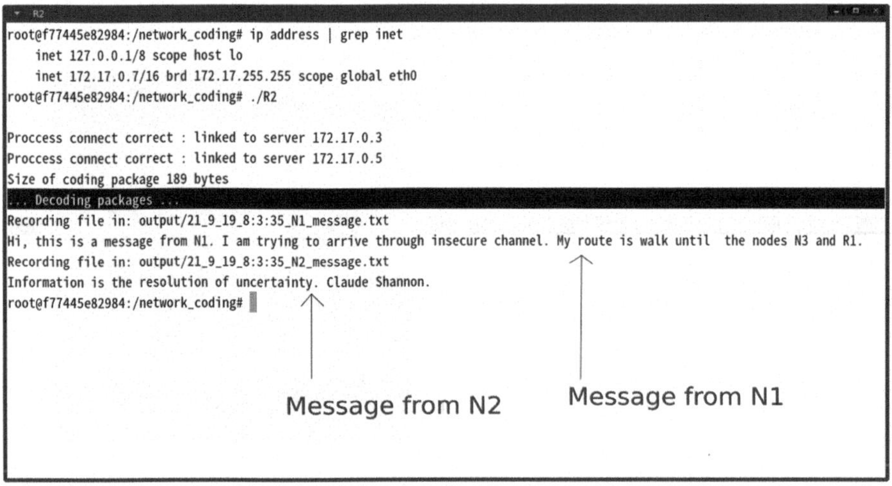

Fig. 18.6 Message recovered in node R2

Fig. 18.7 Case 2

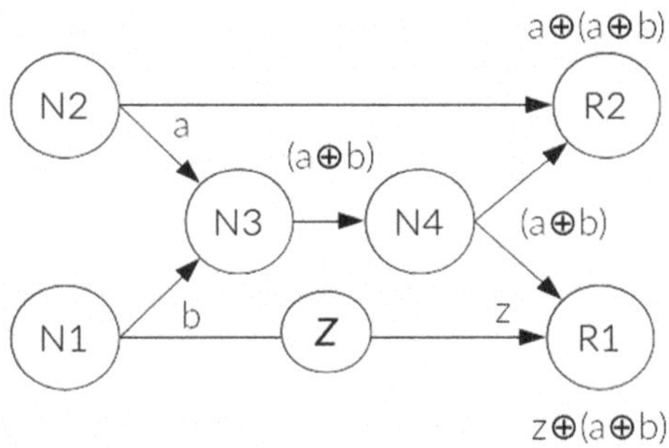

In Case 2 (Fig. 18.7), an attacker sniffs the communication between node N1 and node R1, where node Z represents the pollution attack. Node R1 is seriously affected, because it cannot recuperate the original package from both sources. However, observe that the network may work normally with rest of the nodes, because the process of coding packages operates with normality in node N3.

Now for this scenario, the node Z intercepts communication from node N1 to node R1. Therefore, in R1 the authentic message is lost, receiving instead a contaminated packet. Under a pollution attack process, this case becomes the best case, because it only affects one of the recipients. The other nodes operate correctly. Figure 18.8 shows an example of pollution attack in link between node N1 and node R1. In this example, N1 wants to send an image. But in decoding time, the buffer presents errors. The other packet (from N2) was decoded without difficulties.

Another possibility is shown in Fig. 18.9. Here, pollution attack starts from source N2. The attacker Z cannot only inject corrupt flow into the buffers, also can sniff the communication between R2 and N3 with N2. This produces high impact

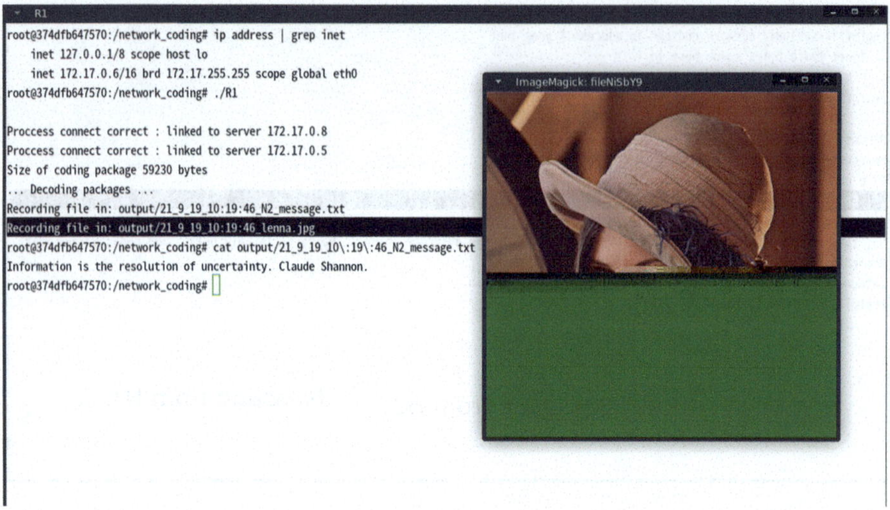

Fig. 18.8 Example of a contaminated buffer

Fig. 18.9 Case 3

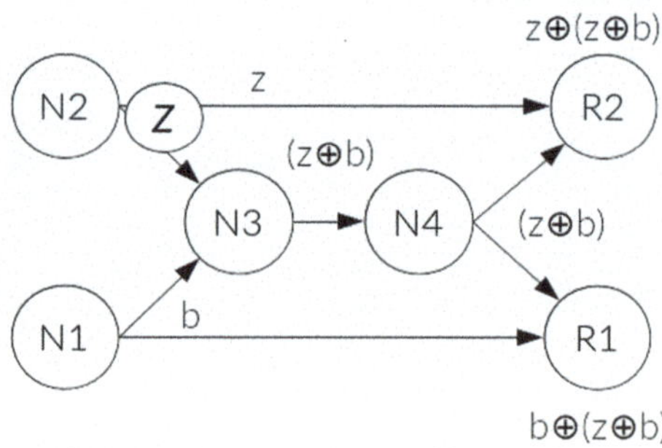

in the network, because network coding is not present correctly, and the network infrastructure is unable to operate properly.

Finally, case 4 in Fig. 18.10 represents a high and dangerous impact to the schema. Here, all packets were intercepted, and the intermediate nodes are propagating the infection to the sinks.

In this case, the intermediate node N3 is intercepted by Z, which injects a contaminated packet in N4. The latter propagates the attack on receivers. Under this attack the following happens. Node Z can spy on the communication of the entire network and it is possible that it injects garbage into the communication channels. Figure 18.11 shows messages sent from the nodes N1 and N2 are intercepted by node Z. Although both messages are received correctly by node Z, this node can inject garbage and send a contaminated message to nodes R1 and R2 across the N4. Figure 18.12 shows how the sink R1 correctly receives packet A (N1_message.txt) directly from the source. While the packet (encoded) received from the node

Fig. 18.10 Case 4

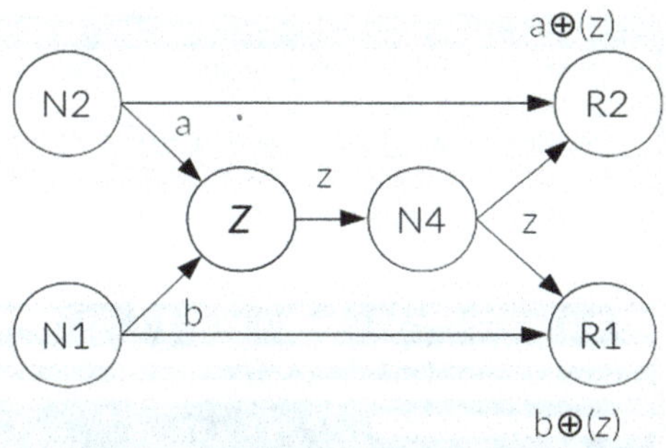

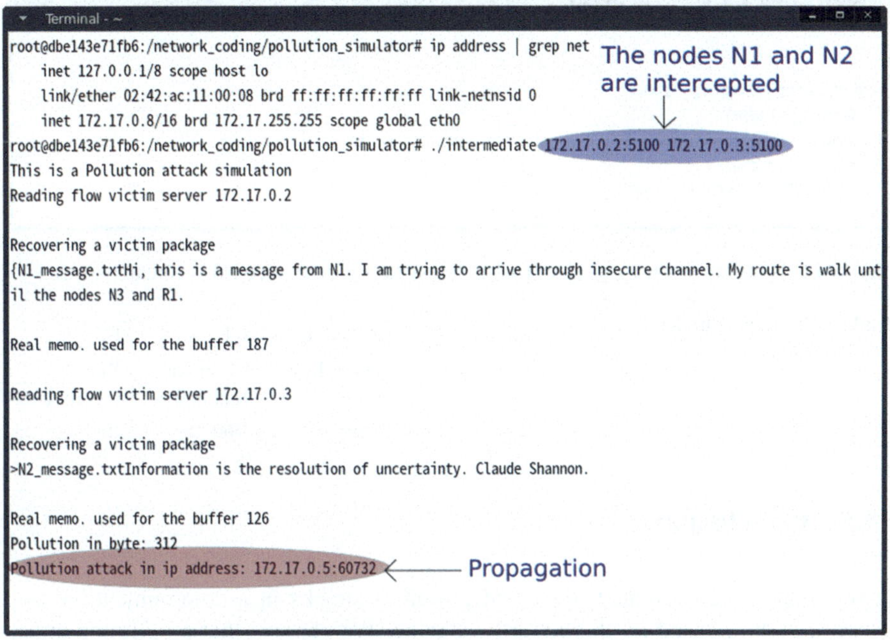

Fig. 18.11 Node Z creates a pollution attack, which is propagated to N4

N3 is contaminated. Consequently the node R1 cannot read the packet B (N2 message.text). The same occurs with the packet A destined to R2.

Table 18.1 summarizes these cases for the sinks R1 and R2, depending where attack occurred. “Complete” is the stage when the messages from the sources were successfully received.

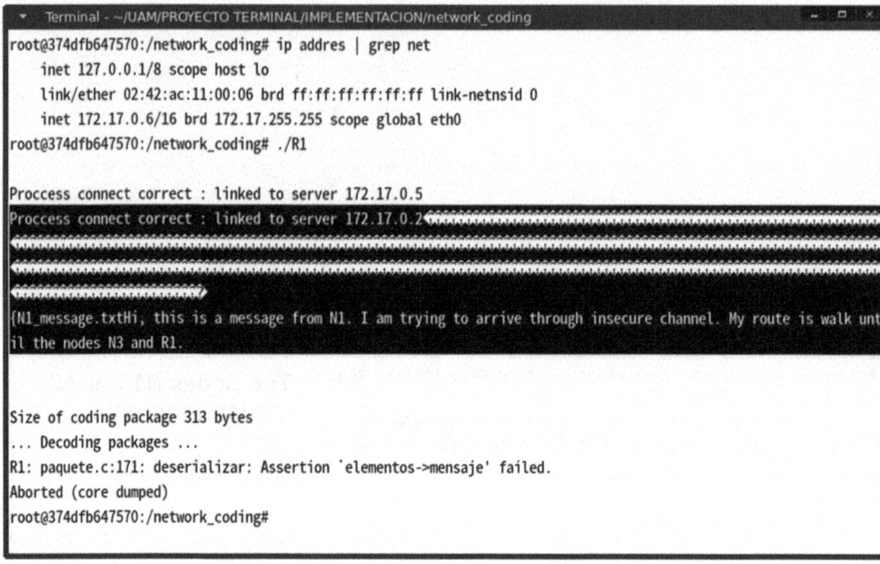

Fig. 18.12 Node R1 correctly receives packet A, but packet from node N3 is contaminated

Table 18.1 Impact under pollution attack

Attack in	R1	R2
N2 with R2	Complete	Bad
Only N1	Incomplete	Incomplete
Only N3	Bad	Bad

18.5 Conclusions

In this paper, we evaluated impact of pollution attacks in a communication with network coding based on the butterfly scheme. We observe that a pollution attack has a different impact in different points in the network. Our investigation is in progress, and as future contribution, we plan to extend our work toward design and implementation of a mechanism to detect packet injection and to mitigate the damage in schemes with network coding.

References

1. R. Ahlswede, N. Cai, S.-Y. Li, R.W. Yeung, Network information flow. IEEE Trans. Inform. Theory **46**, 1204–1216 (2000)
2. T. Ho, D. Lun, *Network Coding: An Introduction* (Cambridge University Press, Cambridge, 2008)
3. D. Lucani, M. Pedersen, D. Ruano, C. Sorensen, F. Fitzek, J. Heide, O. Geil, Fulcrum network codes: A code for fluid allocation of complexity. CoRR, vol. abs/1404.6620, (2015)

4. F.A. Lopez-Fuentes, C. Cabrera Medina, Network coding for streaming video over p2p networks, in *2013 IEEE International Symposium on Multimedia (ISM)*, Anaheim, CA, 9–11 December 2013, pp. 329–332
5. P. Zhang, C. Lin, *Security in Network Coding*, 1st edn. (Springer, Switzerland, 2016)
6. X. Wu, Y. Xu, C. Yuen, L. Xiang, A tag encoding scheme against pollution attack to linear network coding. IEEE Trans. Parallel Distrib. Syst. **25**, 33–42 (2014)
7. W. Xiaohu, X. Yinlong, Y. Chau, X. Liping, A tag encoding scheme against pollution attack to linear network coding. IEEE Trans Parallel Distrib. Syst. **25**(1), 33–42 (2014)
8. L. Guangjun, Security analysis and improvement of a tag encoding authentication scheme for network coding. Wuhan Univ. J. Nat. Sci. **21**(5), 394–398 (2016)
9. J. Vilela, L. Lima, J. Barros, Lightweight Security for Network Coding, in *Proceedings of IEEE International Conference on Communications IEEE ICC'08*, Beijing, China, 19–23 May 2008, pp. 1750–1754

Chapter 19
An Application of Hadamard Transform to Test Stream Ciphers

Guillermo Sosa-Gómez, Omar Rojas, and Octavio Páez-Osuna

Contents

19.1 Introduction 255
19.2 Hadamard Transform 256
19.3 Development 256
19.4 Example 259
19.5 Conclusions 262
References 262

19.1 Introduction

When designing stream ciphers, a number of properties must be taken into account besides generating *randomly looking* bit sequences [4]; the generated sequence must present the most unpredictable behavior possible, that is, given a fraction of the sequence, it should not be possible to predict the rest, either before or after the given subsequence.

In this chapter, we develop theoretical statistical attacks by searching for auto-correlations in the output bits of stream ciphers. The term *cryptanalysis* is also used to refer to any attempt to circumvent the security of different types of algorithms and cryptographic protocols in general, and not just encryption [7]. Although the objective has always been the same, i.e., totally breaking the cryptosystem, the methods and techniques of cryptanalysis have changed drastically throughout the history of cryptography, adapting to a growing cryptographic complexity, which

G. Sosa-Gómez (✉) · O. Rojas
Universidad Panamericana, Escuela de Ciencias Económicas y Empresariales, Zapopan, Jalisco, México
e-mail: gsosag@up.edu.mx; orojas@up.edu.mx

O. Páez-Osuna
Ronin Institute for Independent Scholarship, Montclair, NJ, USA
e-mail: octavio.paezosuna@ronininstitute.org

P. Vasant et al. (eds.), *Data Analysis and Optimization for Engineering and Computing Problems*, EAI/Springer Innovations in Communication and Computing,
https://doi.org/10.1007/978-3-030-48149-0_19

ranges from the pen and paper methods of the past, through machines like Enigma to the systems based on modern computers and other electronic devices.

To overcome the complexities of modern-day cryptography, we must resort to advanced mathematics and algorithms. A very useful mathematical technique to reduce the complexity of a problem is that of changing its domain by means of a transformation. A transform represents the change from one domain to another, and with the right properties, it reduces the complexity of the given task. In the context of our problem, the cryptanalysis of stream ciphers, the Hadamard transform will prove to be useful.

19.2 Hadamard Transform

The Hadamard transform is perhaps the best known of non-sinusoidal orthogonal transformation. The Hadamard transform has gained prominence in applications in the processing of digital signals because it only uses sums and subtractions to compute. Therefore, its implementation in hardware is very efficient [3].

Let $\mathbb{V}_n$ be the vector space of dimension n over the binary Galois field $\mathbb{F}_2$. For two vectors a,$b \in \mathbb{F}_2^n$, we define the scalar product $a \cdot b = (a_1 b_1 \oplus \ldots \oplus a_n b_n)$ and the sum $a \oplus b = (a_1 \oplus b_1, \ldots, a_n \oplus b_n)$, where the product and the sum $\oplus$ (called XOR) are over $\mathbb{F}_2$. We recommend the articles of Bernasconi et al. [1], Walmsley [9], and Pommerening [6] for more on this topic.

Definition 19.2.1 The Hadamard transform of a function f on $\mathbb{V}_n$ (with the values of f taken to be real numbers) is the mapping $H(f) : \mathbb{V}_n \to \mathbb{R}$, defined by

$$H(f)(h) = \frac{1}{N} \sum_{x \in \mathbb{V}_n} f(x)(-1)^{h \cdot x}. \tag{19.1}$$

It can be verified that a direct calculation of the complete Hadamard spectrum using previous definition implies a complexity of N^2 steps, with $N = 2^n$. However, there is a quick procedure to calculate the Hadamard transform that can be computed with only $Nlog(N)$ steps, using the concept of a butterfly diagram (see [2]).

19.3 Development

Cryptanalysis is the study of methods to reveal the meaning of encrypted information, without access to the secret key. Typically, this translates into obtaining the key used to encrypt the information. In nontechnical terms, this practice is known as breaking or forcing the cryptosystem; we will not discuss the details.

Despite that the aim has always been the same, the methods and techniques of cryptanalysis have changed drastically throughout the history of cryptography, adapting to an increasing cryptographic complexity. The methods of cryptanalysis have also changed as it is no longer possible to have unlimited success in breaking a cryptosystem, and there is a hierarchical classification of what constitutes an attack in practice (see [5]).

In the domain of random sequences, there are those generated by intrinsically random physical processes, i.e., based on the assumption of the existence of random processes in nature. In many applications, it is necessary to have the same sequence (apparently random) in two different experiments, so it is necessary to use irreproducible deterministic algorithms. Sequences produced by such algorithms are called pseudo-random. Pseudo-random sequences are used in various environments related to telecommunications [8].

One of the most important steps in cryptanalysis is to establish correspondences between the elements of the output with probable clear text. The main task is to find ways of facilitating the computation of a probability that relates several terms of an output, and determine the likelihood to decipher the message or parts of it.

Let $Z = \{z_1, z_2, \ldots, z_p\}$ be the output of a binary generator, and $z^j = (z_j, z_{j+1}, \ldots, z_{j+k-1}) \in \mathbb{V}_k$ a rolling sequence of Z for all $j \in J = \{1, \ldots, p-k+1\}$ and $\xi = (\xi_1, \ldots, \xi_k) \in \mathbb{V}_k$ for a fixed k.

In our analysis, we want to find the probability $P\{z^j \cdot \xi = 0\}$ for $\xi \neq 0$. Our main goal is the computation of this probability, and thus we start by defining $n_0 =| \{j \in J : z^j \cdot \xi = 0\} |$, and $n_1 =| \{j \in J : z^j \cdot \xi = 1\} |$, such that $n_0+n_1 = N := p-k+1$.

Then, by a classical result in probability, it follows that

$$P\{z^j \cdot \xi = 0\} = \frac{n_0}{N}, \tag{19.2}$$

for all $j \in J$. This probability depends of ξ.

Now, before we proceed to apply certain transformations, we let

$$\frac{1}{N}\sum_{i=1}^{N}(-1)^{z^j \cdot \xi} = \frac{1}{N}\left(\sum_{z^j \cdot \xi = 0}(-1)^{z^j \cdot \xi} + \sum_{z^j \cdot \xi = 1}(-1)^{z^j \cdot \xi}\right) = \frac{1}{N}(n_0 - n_1) = \Delta_\xi. \tag{19.3}$$

Then,

$$\Delta_\xi = \frac{n_0 - n_1}{n_0 + n_1}. \tag{19.4}$$

The problem is, given Δ_ξ, to find the values of n_0 and n_1, which leads to a Diophantine equation that due to the conditions of the previous variables makes it necessary to resort to resources such as Kronecker's Theorem, which makes the solution of the problem very complicated. The idea then is to use a transform that as it is known represents the change from one domain to another and that due to its properties, it reduces the complexity of mathematical problems. This type of tool has been very useful and fundamental in solving problems in different fields, and of diverse nature. The Hadamard transform allows us to solve the problem posed.

Thus, we transform the problem using the Hadamard transform. It turns out that by finding a relationship between Δ_ξ, n_0, and n_1:

$$1 + \Delta_\xi = 1 + \frac{n_0 - n_1}{n_0 + n_1} = \frac{2n_0}{n_0 + n_1}. \tag{19.5}$$

Then, $\frac{1+\Delta_\xi}{2} = \frac{n_0}{N}$.

The probability that we want to find is

$$P\{z^j \cdot \xi = 0\} = \frac{n_0}{N} = \frac{1 + \Delta_\xi}{2}, \tag{19.6}$$

reducing the problem to that of finding Δ_ξ.

Sequence

Suppose now that we obtain an output sequence Z, we want to count the repetitions of z^j, for this we build the set $\alpha = \{\alpha^l \in \mathbb{V}_k\}$, for $l = 1, \ldots, 2^k$ and we use Alg. 6.

```
Data: z^j
Result: Counter z^j
for l = 1 to 2^k do
    n_l = 0;
    for j = 1 to N do
        if z^j = α^l then
            n_l = n_l + 1;
            break;
        end
    end
end
```

Algorithm 6: Counter

Then we compute the following:

$$\Delta_\xi = \frac{1}{N} \sum_{j=1}^{N} (-1)^{z^j \cdot \xi} = \frac{1}{N} \sum_{\alpha^l \in \alpha} (-1)^{\alpha^l \cdot \xi} \cdot n_l, \tag{19.7}$$

for all $\xi \in \mathbb{V}_k$. Each z^j is mapped to $y^j \in \mathbb{N}$ such that

$$y^j = \sum_{i=0}^{k-1} z_i^j \cdot 2^i, \tag{19.8}$$

for all $j \in J$, then we let $\Pi[y^j] = \Pi[y^j] + 1$, this is a counter for the number of y^j's such that $n_l = \Pi[l]$.

Now let $\hat{n}_\xi = \sum_{\alpha^l \in \alpha} (-1)^{\alpha^l \cdot \xi} n_l$. Then to compute $\{\hat{n}_\xi, \xi \in \mathbb{V}_k\}$ and solve (3), we will use the discrete Hadamard transform. If we write $n_l = f(x)$ and to $(h, x) = (\alpha^l, \xi)$, then the transform can be written as

$$H(n_l)(\xi) = \frac{1}{N} \sum_{\alpha^l \in \alpha} n_l (-1)^{\alpha^l \cdot \xi}, \tag{19.9}$$

arriving to

$$\Delta_\xi = \frac{\hat{n}_\xi}{N} = H(n_l)(\xi), \tag{19.10}$$

which is a less cumbersome computation of Δ_ξ. Once Δ_ξ is obtained, we can determine if there are distinguished elements within $\mathbb{V}_k$ that can help us characterize certain parts of the output sequence Z or in this case by finding a certain relationship between the elements of the sequence.

19.4 Example

In this example, we illustrate our methods by analyzing the shrinking generator depicted in Fig. 19.1. We use LSFR of length 7 A_1 with primitive polynomial

$$z^7 + z^4 + 1,$$

LFSR of length 6 A_2 with primitive polynomial

$$z^6 + z^4 + z^3 + z + 1.$$

Both LFSR are initialized with a random binary sequence of the respective lengths, say $(a_0, \ldots, a_6)$ and $(su_0, \ldots, su_5)$. Let $n_0 = 7$, $n_1 = 2^{n_0} - 1 = 127$, $n_2 = 6$, $n_3 = 2^{n_1} - 1 = 63$. Define $a_{n_1+n_0} := 0$ and $a_{i+n_0} := (a_{i+4} + a_i) \mod 2$ for $i = 0, \ldots, 3000n_1$, let $su_{n_1+n_3} := 0$ and $su_i := (su_{i+4} + su_{i+3} + su_{i+1} + su_i) \mod 2$ for $i = 0, \ldots, 1300n_3$. The aforementioned sequence is generated by the shrinking generator Alg. 7.

Result: z the compressed output
l=0;
for $i = 0$ *to* $rows(su) - 1$ **do**
  if $su_i = l$ **then**
    $z_l = a_i$;
    $l = l + 1$;
  end
  break if $l > rows(a) - 1$;
end

Algorithm 7: The compression algorithm

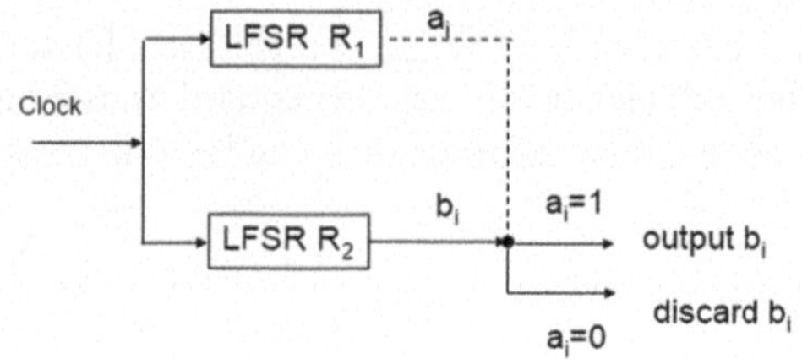

Fig. 19.1 Schematic for generator used on step 1

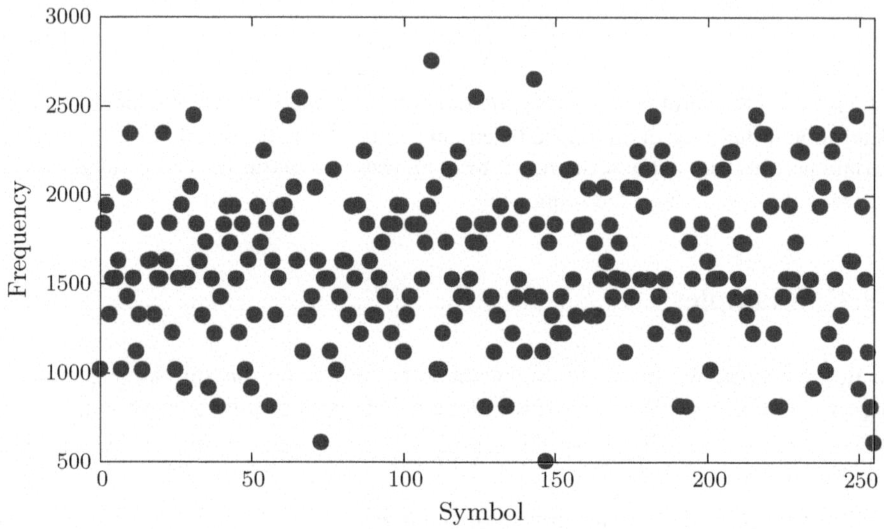

Fig. 19.2 Raw frequencies of symbols in the sample sequence

We now apply the compressor Algorithm 7 to the output sequences $\{a_i\}$ and $\{su_i\}$, let $z = Compressor(a, su)$.

We organize the output Z sequentially in blocks of length $k = 8$, and interpret these blocks as the sequences z^j. It has therefore length $p = 41600$. Each z^j is the binary expression of an integer in the interval $[0, 2^8 - 1]$. We store their frequencies in vector Π of length 2^8; i.e., $\Pi_i = |\{r | i = \sum_{j=0}^{7} z_{8r+j} 2^j, r = 0, \ldots, 41600\}|$.

Figure 19.2 shows the frequency graph of symbols between 0 and 255 of one experiment. As you can see, it is very hard to single out a particular element. We now resort to Hadamard matrices to transform our problem. Recall that we may construct Hadamard matrices recursively, using the Kronecker product $\otimes$, starting with

$$H_1 := \begin{pmatrix} 1 & 1 \\ 1 & -1 \end{pmatrix}. \tag{19.11}$$

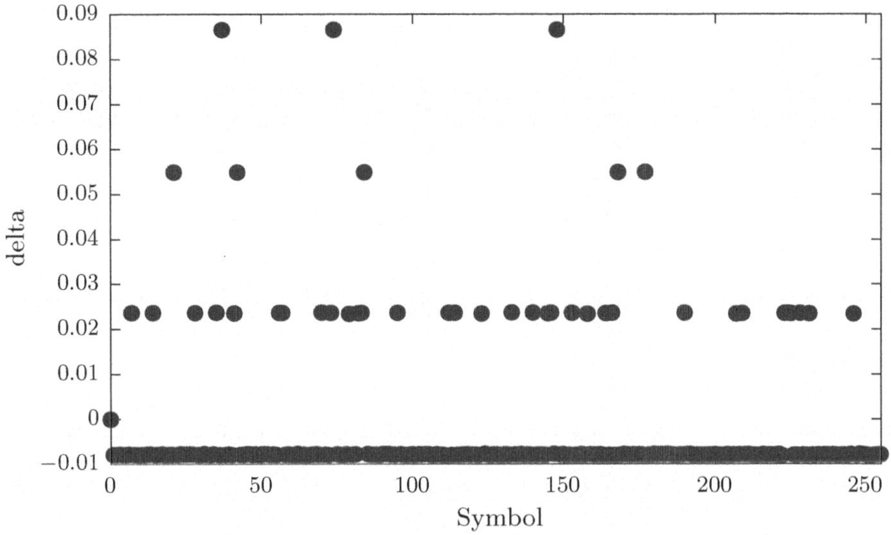

Fig. 19.3 Probability representation of sample frequencies. After transforming distinguished elements are revealed

We proceed to construct

$$H_2 = H_1 \otimes H_1 = \begin{pmatrix} 1 & 1 & 1 & 1 \\ 1 & -1 & 1 & -1 \\ 1 & 1 & -1 & -1 \\ 1 & -1 & -1 & 1 \end{pmatrix} \tag{19.12}$$

Then $H_4 = H_2 \otimes H_2$, and finally $H_8 = H_4 \otimes H_4$, which is the transform we need for our problem. In our example, $N = 41592$ and define $\delta := \frac{H_8 \cdot \Pi}{N}$ and $\delta_0 := 0$.

In this experiment we find that $max(\delta) = 0.087$, corresponding to the 37-th entry. Now $37_{10} = 100101_2$, and with probability $\frac{1+\delta_{37}}{2} = 0.544$, which implies that, with high probability, 37 is a distinguished element from the output. Indeed, we may verify this by computing

$$\frac{\sum_{i=0}^{N}(-1)^{z_i+z_{i+3}+z_{i+5}}}{N} = 0.087 = \delta_{37}, \tag{19.13}$$

which tells us that our computation is correct (Fig. 19.3).

19.5 Conclusions

We presented a method for the cryptanalysis of pseudo-random generators in a stream cipher. The method relies on the discrete Hadamard transform to map the problem from a statistical setting to a probabilistic domain, in which it is possible, with high probability, to single out elements of the output and pair them with their corresponding clear texts. We included an example to illustrate the simplicity of our method and its accuracy.

References

1. A. Bernasconi, B. Codenotti, J. Simon, On the fourier analysis of boolean functions. 1–24 (1996, preprint)
2. J.W. Cooley, J.W. Tukey, An algorithm for the machine calculation of complex Fourier series. Math. Comput. **19**(90), 297 (1965). https://doi.org/10.2307/2003354
3. K.J. Horadam, Hadamard matrices and their applications, in *Hadamard Matrices Their Applications* (Princeton University Press, Princeton, 2012), pp. 1–263. https://doi.org/10.1515/9781400842902
4. M.S. Ibáñez, R.G. Díaz, *Generación y análisis de Secuencias Pseudoaleatorias* (UPC, Barcelona, 1999)
5. J. Katz, Y. Lindell, *Introduction to Modern Cryptography* (Chapman and Hall/CRC, Boca Raton, 2014)
6. K. Pommerening, *Fourier Analysis of Boolean Maps - a Tutorial* (Gutenberg University, Preprint, 2005)
7. B. Schneier, A self-study course in block-cipher cryptanalysis. Cryptologia **24**(1), 18–33 (2000)
8. J. Seberry, B.J. Wysocki, T.A. Wysocki, On some applications of Hadamard matrices. Metrika **62**, 221–239 (2005). https://doi.org/10.1007/s00184-005-0415-y
9. W.M. Walmsley, Walsh functions, transforms and their applications. Electron. Eng. **46**(556), 63–75 (1974)

Chapter 20
A Comparison of Speech-to-Speech Neural Network Methodologies for Digit Pronunciation

Manuel Quintana and Miguel Bernal

Contents

20.1 Introduction 263
20.2 Preliminaries 265
20.3 The First-Grammar-Then-Voice Approach 268
20.4 The First-Voice-Then-Grammar Approach 270
20.5 Conclusions 272
References 272

20.1 Introduction

Motivation This work is a first step toward the investigation of voice-to-voice methodologies, which are scarce in natural language processing (NLP) reports. Our contention is that voice-to-voice methodologies better fit the way a human being develops the ability to speak by supervised learning and implicit adaptation, instead of explicit grammar-oriented approaches, which come much later in the cognitive evolution of a human being. By setting up audio digit recognition for identifying a digit (in Spanish) in two ways: the first-grammar-then-voice (FGTV) approach and the first-voice-then-grammar (FVTG) one, we intend to prove our contention.

Problem Statement Compare the methodologies corresponding to the FGTV approach (already explored in the literature) and the proposed FVTG solution in a very elementary problem with the most basic neural-network tools.

State of the Art NLP comprises a lot of areas, depending on where the focus is made on signal processing [1], human-machine interface [2], or artificial intelligence [3]. Two extreme approaches can be identified: on the one hand, statistical tools are usually applied when the emphasis is made on signal treatment without any higher-

M. Quintana and M. Bernal (✉)
Department of Electrical and Electronics Engineering, Sonora Institute of Technology, Ciudad Obregon, Mexico
e-mail: miguel.bernal@itson.edu.mx

P. Vasant et al. (eds.), *Data Analysis and Optimization for Engineering and Computing Problems*, EAI/Springer Innovations in Communication and Computing,
https://doi.org/10.1007/978-3-030-48149-0_20

level interpretation; on the other hand, purely conceptual handling of syntactic units and semantics, which disregards implementation details in order to better relate with linguistic communities [4]. An in-between approach is that based on deep neural networks as these biologically inspired models have been long ago related with statistical analysis [5], while being capable of subsuming large learning tasks in more conceptual units [6].

In the last decades, neural networks have been well established as a useful tool to perform learning tasks by embedding highly complex, nonlinear, and parallel information into weights that mimic brain synapsis [7]. Three main landmarks can be distinguished in their evolution: a modest origin when the perceptron was nothing else than a linear filter [8], the success of the multilayer perceptron and the backpropagation algorithm for heuristic training [9], and, finally, the surge of deep learning techniques via convolutional, recurrent, and recursive neural networks [10]. The latter framework has already made a significant advance in computer vision and pattern recognition by yielding shallow models such as support vector machines (SVM) to neural networks trained with dense vector representations [11].

Within the field of NLP, neural networks in all their varieties have played a central role [12]: convolutional neural networks (CNNs) have been suitable for feature extraction [13] and data mining [14]; recurrent neural networks (RNN) can be found in word- and sentence-level classification [15, 16]; recursive neural networks (VNNs) have been successfully employed for parsing [17]. In all these tasks, multiple processing layers on which different deep learning algorithms can be applied, have been used to learn hierarchical data.

An important task within the NLP area is that of automated translation, which usually involves feature analysis (based on neural networks as well as Markov chains, for which software tools such as HTK or SPHINX are available [18]), unit detection (such as phonemes, [19]), syntactic analysis (for example, for word validation [20]), and proper translation via a looking-up procedure in a database. Remarkably, to the best of our knowledge, voice-to-voice translation by direct learning is not present in the literature with few exceptions for classification [21–24].

Contribution A comparison between voice-to-voice methodology for translation against grammar-based approaches is provided in the simplest neural network architecture (multilayer perceptron) for a very common problem (digit pronunciation). The results show that, even without the improvements of deep learning, it is possible to provide translation as an encoder, which better fits the way a human being learns to speak: this may save training time and avoids the overspecialization of tasks in the grammar-based solutions.

Organization This work is organized as follows: Section 20.2 makes an elementary summary of what we have to know about audio signals and preprocessing while describing the neural network and learning algorithm employed to perform the learning tasks: they have been chosen as elementary as possible to facilitate reproducibility; Sect. 20.3 details the experimental setup for the FGTV approach, which in this case simply consists on identifying the digits to pick up the right class for pronunciation and/or translation; Sect. 20.4 describes the second approach, which intends to ignore the grammar structure by simply reproducing the different examples, which is equivalent to teach pronunciation to the network regardless of the meaning; conclusions are gathered in Sect. 20.5.

20.2 Preliminaries

In this section, we discuss the way audio signals are obtained, normalized, and the sort of neural network we are employing. It should be kept in mind that every choice has been made as simple as possible as to guarantee reproducibility of the results here provided.

Audio Signals

When recorded via microphones or other electronic devices, audio signals are usually displayed as a time function resulting from the vibrations of a membrane; its intensity (see Fig. 20.1, which corresponds to the Spanish word for zero) roughly corresponds to the volume and its variations have frequency content that can be further analyzed via a spectrogram such as that in Fig. 20.2. A spectrogram combines information in the temporal and frequency domains: its horizontal axis represents time while the vertical one corresponds to frequency; the latter is usually split in bands of 1000 Hz each, and ranges from 0 to 8000 Hz, which is the highest frequency to be found in human voice [1], though that in Fig. 20.2 is limited to 5000 Hz. The darker an area in the spectrogram is, the higher the energy of those frequencies at that time; when these peaks are above a certain level they are identified as formants of the signal, which play an important role in phonetic analysis [25] (they are identified by bullets in Fig. 20.2).

As with many other works on audio analysis, the tool we are employing to capture the audio signals is Praat, a computer program from the University of Amsterdam for analysis and manipulation of speech [26]; the tool is well beyond

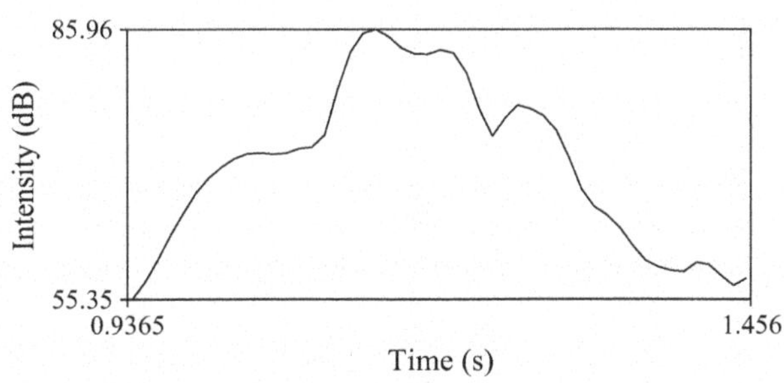

Fig. 20.1 An intensity signal: time VS intensity by RMS analysis

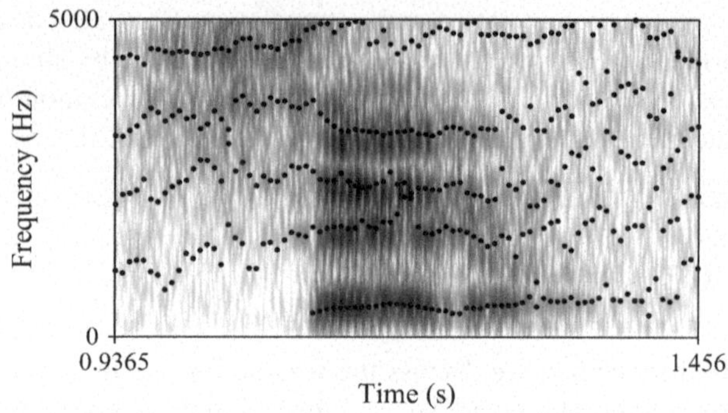

Fig. 20.2 A spectrogram: time VS frequencies. Formants are highlighted as bullets

the needs of this report as spectral, pitch, formant, and intensity analysis, among others, can be performed with it. Tutorials are available for phonetics, learning, and statistics [27].

Clearly, a deep understanding of audio analysis is out of the scope of this work as we are only concerned with capturing audio signals in any form so that they can be preprocessed and fed to a neural network for pattern recognition. It is customary to use a grid of the spectrogram to vectorize its lighter/darker contents for neural network processing [28, 29]; this approach is closely linked to image processing as proven in [30]. In this work, we employ the intensity signal as the digits spoken in Spanish present distinctive features that, in our opinion, do not require going further to spectral analysis, though of course a grid of the spectrogram can be employed instead with minimum changes; a similar approach can be found in [31] for the Portuguese language.

Preprocessing of the Audio Signals

As digits may be spoken in a variety of volumes and durations, they require a preprocessing in order to make any sample signal have the same wide and maximum height. A number of methods are available for neural-network-oriented preprocessing of audio signals [32–35], but, as we want to keep the structures and algorithms as simple as possible to illustrate our hypothesis, we will proceed with an ad hoc regularization of the samples that simply shrinks/expands audio signals to a "time" window of 1000 units (these are no longer seconds) and every magnitude proportionally raised/lowered to a fraction of the maximum height fixed at 1 units (these are no longer decibels).

The formula for such changes is a simple interpolation formula for the horizontal scaling (which can be implemented in MATLAB^{TM} via interp1), while the vertical adjustment is given afterwards by an ordinary rule of three

$$h_{new} = h_{current}/h_{max}, \tag{20.1}$$

where $h_{current}$ is the height of the current point, h_{max} it the maximum height of the whole sample, and h_{new} is the scaled height. After these changes, every audio sample signal will consist of 1000 points with magnitudes in the range [0, 1]. This is the vector that will be feed for neural network training.

The Neural Network and the Learning Algorithm

A standard 1000-input fully connected 2-hidden layer neural network with 70 and 25 neurons and 10 outputs will be employed for the FGTV approach; the number of outputs corresponds to the 10 different digits to be identified in the FGTV approach, each of them is then pronounced in a standard form. Figure 20.3 is a simplified scheme of this architecture. On the other hand, for the FVTG approach a 1000-input fully connected 2-hidden layer neural network with 70-70 neurons and 1000 outputs is going to be trained to reproduce with its "own voice" the corresponding digit. Obviously, the architecture is very similar to that in Fig. 20.3 with the corresponding changes in the number of neurons and outputs.

The training algorithm is a standard backpropagation with the following choices [7]:

1. Initialization: Weights in the neural network are randomly initialized in a range $2\epsilon \times [0, 1]$, $\epsilon \approx 0.1$.
2. Training samples: Each sample has the standardized intensity curve form; it consists of 1000 points whose values are in the range [0, 1]. At each epoch all the samples are presented in a random order.
3. Forward computation: Induced local fields are calculated in a standard way, each neuron includes a bias term of 0.8 whose weights are initialized and updated in the same fashion as the others. Activation functions are all logistic functions

$$\phi(v) = 1/(1 + \exp(-v)). \tag{20.2}$$

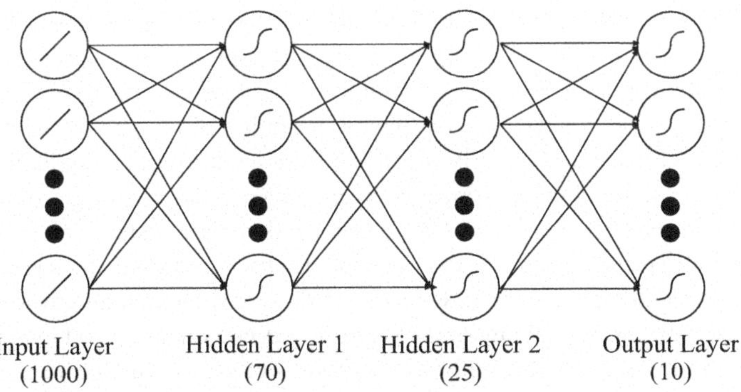

Fig. 20.3 The neural network architecture: fully connected multilayer perceptron

4. Backward computation: Each weight is updated with a standard local gradient approach, including a momentum term, i.e.,

$$w_{ji}^{(l)}(n+1) = w_{ji}^{(l)}(n) + \alpha \Delta w_{ji}^{(l)}(n-1) + \eta \delta_j^{(l)}(n) y_i^{(l-1)}(n), \tag{20.3}$$

where n represents the current instant, w_{ji} the weight from the i-th neuron/entry in the $(l-1)$-th layer to the j-th neuron/output in the l-th, α is a momentum parameter, η is the learning rate, $\delta_j^{(l)}$ is the local gradient, and $y_i^{(l-1)}$ the i-th output of the previous layer. Local gradients are calculated in standard fashion too; it is omitted for brevity; the interested reader is referred to [7].

20.3 The First-Grammar-Then-Voice Approach

As stated before, the normalized intensity signal of each sample is employed for training; they are displayed in Fig. 20.4. Normalization guarantees they consist all in 1000 points and the maximum height is 1. Clearly, the signals in Fig. 20.4 belonging to each digit share some common characteristics, though some of them are more inclined to have important variations, namely, the digits 4 and 8, the latter due to pronunciation differences of "ch" in Spanish "ocho": some of them phonetically closer to "sh" than to "ch" in Northwest Mexico.

For pronunciation/translation of the digits, the FGTV approach does not care for learning how to pronounce, but instead it identifies the digit and then chooses among a variety of standardized sounds to pronounce or translate (or any other speech treatment for that matter). It is therefore enough to have 10 outputs in the neural network, which has been tuned with the following parameters: $\alpha = 0.1$, $\eta(0) = 0.12$ with progressive linear reduction until $\eta(N) = 0.01$, where $N = 100$ is the number of samples per epoch (i.e., 10 samples of each digit).

Three different number of epochs were tested: 500, 1000, and 5000, each of them was run 100 times to obtain the averages in Table 20.1. It is clear that the higher the number of epochs, the better the recognition rate among the training samples; this is not the case for the test samples, which do not belong to the training set. In the latter case, the recognition rate drops significantly, perhaps due to overfitting as small variations in the intensity signal are enough to misclassify the sample. In Fig. 20.5 the error norm $\|e(t)\|$ per epoch is displayed: though it diminishes sharply at first, it does not keep the pace and quickly reaches a limit (with frequent sudden changes).

In summary, if the FGTV approach is employed, the resources of the neural network are employed to identify the *class* of the digit instead of its *shape* for reproduction. Identifying the class is consistent with a grammar-oriented approach, which intends to identify words, classes of words, and their semantic role for speech processing. However, for certain activities it might be beneficial to teach the neural network to *reproduce* instead without directing the efforts of the learning task to grammar tasks: this is shown in the next section.

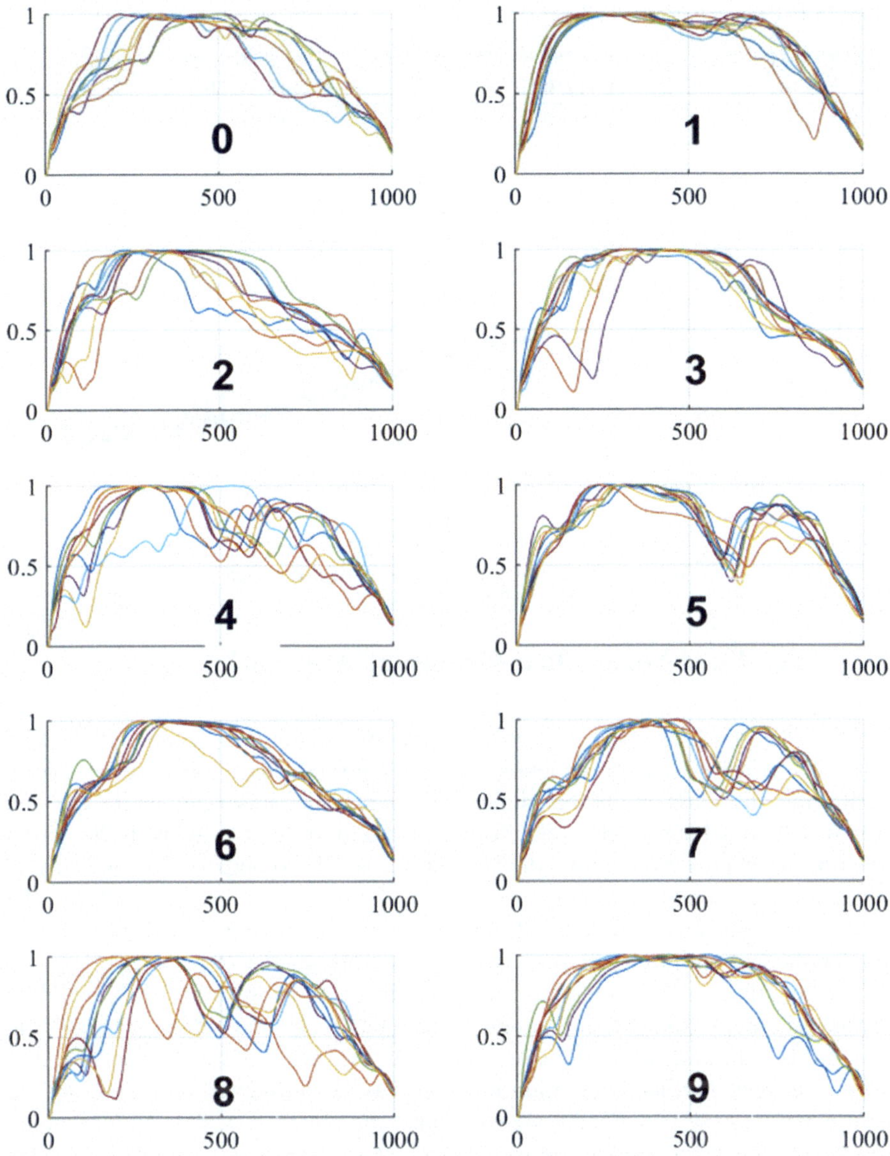

Fig. 20.4 Normalized intensity signals of the digits 0–9

Table 20.1 Recognition rates in the FGTV approach

Number of epochs	Recognition rate (training)	Recognition rate (test)
500	71.41/100	18.35/30
1000	92.59/100	18.01/30
5000	93.28/100	18.37/30

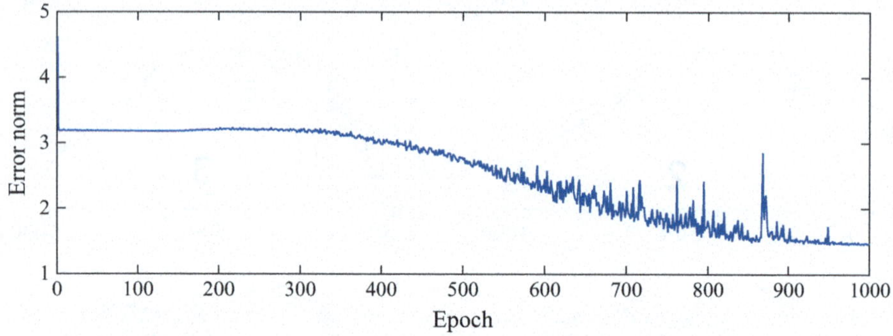

Fig. 20.5 Error norm $\|e(t)\|$ in the First-Grammar-Then-Voice Approach

20.4 The First-Voice-Then-Grammar Approach

The FVTG approach does not care for identifying the digit class, but instead it intends to reproduce the intensity signal for reproduction "on its own voice." Identifying the class might be viewed as a grammar approach, while the one proposed here belongs to the voice-to-voice paradigm. In contrast with the neural network in the previous section, the current one has 1000 outputs, i.e., as many as the normalized input as it is going to work as a *replicator* in the spirit of [36]. The following parameters have been chosen: $\alpha = 0.1$, $\eta(0) = 0.1$ with progressive linear reduction until $\eta(N) = 10e^{-5}$, where $N = 60$ samples (6 per digit).

The neural network training has been performed employing 500, 1000, and 5000 epochs for comparison. All of them present a remarkable reduction of the quadratic error as shown in Fig. 20.6 for 5000 epochs. As the neural network is supposed to work as a replicator, its recognition rate can only be considered by measuring the error norm of a particular input signal against the obtained output. To have some measure of this error, another 60 test signals were applied to the trained network; for each of them, the average error is calculated as follows, where $d(i)$ and $o(i)$ are the desired and obtained values per point:

$$e_j = \left(\frac{1}{1000} \sum_{i=1}^{1000} (d(i) - o(i))^2 \right)^{\frac{1}{2}}, \quad j \in \{1, 2, \ldots, 60\}, \tag{20.4}$$

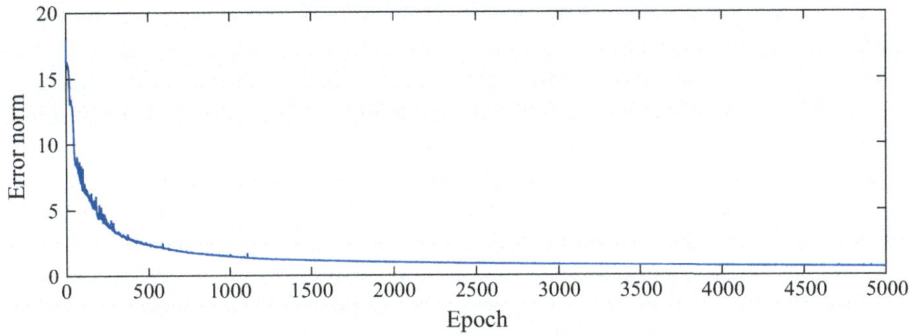

Fig. 20.6 Error norm $\|e(t)\|$ in the First-Voice-Then-Grammar Approach

Table 20.2 Recognition rates in the FVTG approach

Number of epochs	Averaged error (20.5) (training)	Averaged error (20.5) (test)
500	0.0242	0.0882
1000	0.0152	0.0760
5000	0.0054	0.0681

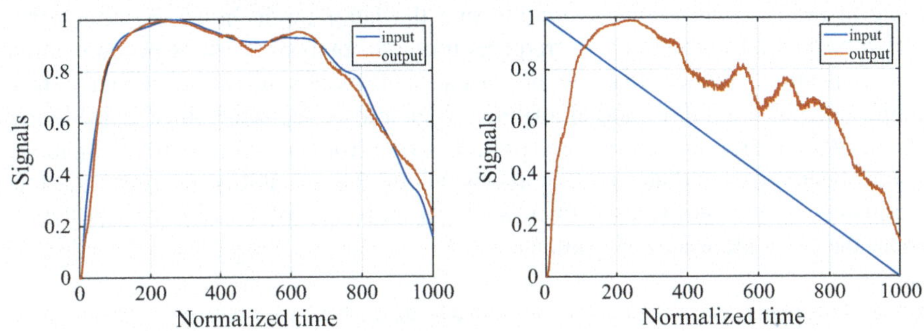

Fig. 20.7 Test signals and their outputs: valid signal (left), invalid signal (right)

$$e_{final} = \frac{1}{60}\sum_{j=1}^{60} e_j. \tag{20.5}$$

The results are displayed in Table 20.2 both for the training and test samples.

Note this is not an identity map because for each unknown entry the neural network picks up the closest representation among those learnt by the network. To see this, consider Fig. 20.7, where 2 input signals outside the training set were applied to one of the 5000-epoch trained neural network. Note that the resulting "voice intensity" closely matches that of the entry in valid numbers (left, corresponding to Spanish "uno," i.e., 1), but does not reproduce signals outside the training (right). Notice also that the output of the trained network has a main signal with a high-frequency low-magnitude noise over it; this noise can be easily eliminated via low-pass filters.

Frequency contents in the intensity signals are embedded as they follow the original raw audio variations whose frequency contents are displayed in a spectrogram (see Fig. 20.2). Thus, reproducing the resulting voice from the neural network in the approach described in this section would imply to lend it a fixed frequency that can be taken using the short-time Fourier transform [1]. Another problem would be to adjust the time length and volume, the first by taking the average length of the raw input signals (once identified) and the second by scaling the intensity. Yet, keep in mind that this work is intended to prove that "in principle" such task is possible, even if more data need to be included for a proper artificial voice (pitch, formants, etc.).

20.5 Conclusions

A comparison between a first-voice-then-grammar methodology for speech processing has been contrasted with the more preferred approach of first-grammar-then-voice in order to prove the contention that the latter might be more suitable for certain tasks that not necessarily require grammar analysis, such as pronunciation and translation. The comparison has been performed with the most elementary neural network structures and algorithms to guarantee reproducibility, considering the problem of digit recognition and pronunciation from an audio source in Spanish. It has been proved that our hypothesis about the suitability of replication is promising, which motivates further work on more complex tasks for translation under the same grammarless grounds.

Acknowledgments This work has been supported by the CONACYT scholarship 725949 and the ITSON PFCE 2019.

References

1. A. Lerch, *An Introduction to Audio Content Analysis: Applications in Signal Processing and Music Informatics* (Wiley-IEEE Press, Hoboken, 2012)
2. O. Poleshchuk, E. Komarov, *Expert Fuzzy Information Processing*, vol. 268 (Springer Science & Business Media, New York, 2011)
3. J. Kacprzyk, S. Zadrozny, Computing with words is an implementable paradigm: fuzzy queries, linguistic data summaries, and natural-language generation. IEEE Trans. Fuzzy Syst. **18**(3), 461–472 (2010)
4. J.P. Carvalho, On the semantics and the use of fuzzy cognitive maps and dynamic cognitive maps in social sciences. Fuzzy Sets Syst. **214**, 6–19 (2013)
5. N.V. Vladimir et al., *Statistical Learning Theory* (Wiley, New York, 1998). Translation J.H. Xu, X.G. Zhang (Publishing House of Electronics Industry, Beijing, 2004)

6. S.K. Pal, S. Mitra, *Neuro-Fuzzy Pattern Recognition: Methods in Soft Computing* (Wiley, New York, 1999)
7. S.S. Haykin, S.S. Haykin, S.S. Haykin, K. Elektroingenieur, S.S. Haykin, *Neural Networks and Learning Machines*, vol. 3 (Pearson Education, Upper Saddle River, 2009)
8. F. Rosenblatt, Principles of neurodynamics. Perceptrons and the theory of brain mechanisms. Cornell Aeronautical Lab Inc., Buffalo, Tech. Rep. (1961)
9. P.J. Werbos et al., Backpropagation through time: what it does and how to do it. Proc. IEEE **78**(10), 1550–1560 (1990)
10. I. Goodfellow, Y. Bengio, A. Courville, *Deep Learning* (MIT Press, Cambridge, 2016)
11. T. Young, D. Hazarika, S. Poria, E. Cambria, Recent trends in deep learning based natural language processing. IEEE Comput. Intell. Mag. **13**(3), 55–75 (2018)
12. S. Bird, E. Klein, E. Loper, *Natural Language Processing with Python: Analyzing Text with the Natural Language Toolkit* (O'Reilly Media, Inc., Sebastopol, 2009)
13. S. Poria, E. Cambria, D. Hazarika, P. Vij, A deeper look into sarcastic tweets using deep convolutional neural networks. Preprint, arXiv:1610.08815 (2016)
14. M. Nafari, C. Weaver, Query2question: translating visualization interaction into natural language. IEEE Trans. Vis. Comput. Graph. **21**(6), 756–769 (2015)
15. G. Lample, M. Ballesteros, S. Subramanian, K. Kawakami, C. Dyer, Neural architectures for named entity recognition. Preprint, arXiv:1603.01360 (2016)
16. X. Wang, Y. Liu, S. Chengjie, B. Wang, X. Wang, Predicting polarities of tweets by composing word embeddings with long short-term memory, in *Proceedings of the 53rd Annual Meeting of the Association for Computational Linguistics and the 7th International Joint Conference on Natural Language Processing (Volume 1: Long Papers)*, vol. 1, 2015, pp. 1343–1353
17. R. Socher, C.C. Lin, C. Manning, A.Y. Ng, Parsing natural scenes and natural language with recursive neural networks, in *Proceedings of the 28th International Conference on Machine Learning (ICML-11)*, 2011, pp. 129–136
18. W. Walker, P. Lamere, P. Kwok, B. Raj, R. Singh, E. Gouvea, P. Wolf, J. Woelfel, Sphinx-4: a flexible open source framework for speech recognition (2004)
19. H. Krupakar, K. Rajvel, B. Bharathi, S.A. Deborah, V. Krishnamurthy, A survey of voice translation methodologies—acoustic dialect decoder, in *2016 International Conference on Information Communication and Embedded Systems (ICICES)* (IEEE, Piscataway, 2016), pp. 1–9
20. J. Su, J. Zeng, D. Xiong, Y. Liu, M. Wang, J. Xie, A hierarchy-to-sequence attentional neural machine translation model. IEEE/ACM Trans. Audio Speech Lang. Process. **26**(3), 623–632 (2018)
21. P.K. Sharma, N. Sahu, A hybrid voice identification system with fuzzy technique and ART2 neural network on BPF technique. J. Commun. Eng. Syst. **8**(3), 1–6 (2018)
22. W.C. Huffman, M. Pappas, System and method for voice-to-voice conversion, 29 Nov 2018, US Patent App. 15/989062
23. N. Lavan, S.E. Merriman, P. Ladwa, L.F. Burston, S. Knight, C. McGettigan, 'Please sort these voice recordings into 2 identities': effects of task instructions on performance in voice sorting studies. Br. J. Psychol. (2019). https://doi.org/10.1111/bjop.12416
24. A. Lieto, D. Moro, F. Devoti, C. Parera, V. Lipari, P. Bestagini, S. Tubaro, "Hello? who am i talking to?" a shallow cnn approach for human vs. bot speech classification, in *ICASSP 2019-2019 IEEE International Conference on Acoustics, Speech and Signal Processing (ICASSP)* (IEEE, Piscataway, 2019), pp. 2577–2581
25. P. Ladefoged, S.F. Disner, *Vowels and Consonants* (Wiley, New York, 2012)
26. W. Styler, Using praat for linguistic research. University of Colorado at Boulder Phonetics Lab (2013)
27. P. Boersma, D. Weenink, Praat, a system for doing phonetics by computer (version 6.0. 28). Institute Of Phonetic Sciences University of Amsterdam (up-to-date version of the manual at http://www.fon.hum.uva.nl/praat/) (2017)
28. B.E. Kingsbury, N. Morgan, S. Greenberg, Robust speech recognition using the modulation spectrogram. Speech Commun. **25**(1–3), 117–132 (1998)

29. P. Khunarsal, C. Lursinsap, T. Raicharoen, Singing voice recognition based on matching of spectrogram pattern, in *2009 International Joint Conference on Neural Networks* (IEEE, Piscataway, 2009), pp. 1595–1599
30. B. Pinkowski, Principal component analysis of speech spectrogram images. Pattern Recognit. **30**(5), 777–787 (1997)
31. D.F. Silva, V.M. de Souza, G.E. Batista, R. Giusti, Spoken digit recognition in portuguese using line spectral frequencies, in *Ibero-American Conference on Artificial Intelligence* (Springer, Berlin, 2012), pp. 241–250
32. L. Muda, M. Begam, I. Elamvazuthi, Voice recognition algorithms using mel frequency cepstral coefficient (MFCC) and dynamic time warping (DTW) techniques. Preprint, arXiv:1003.4083 (2010)
33. K. Choi, D. Joo, J. Kim, Kapre: on-GPU audio preprocessing layers for a quick implementation of deep neural network models with keras. Preprint, arXiv:1706.05781 (2017)
34. Y. Han, J. Park, K. Lee, Convolutional neural networks with binaural representations and background subtraction for acoustic scene classification, in *The Detection and Classification of Acoustic Scenes and Events (DCASE)* (2017), pp. 1–5
35. K. Choi, G. Fazekas, M. Sandler, K. Cho, A comparison of audio signal preprocessing methods for deep neural networks on music tagging, in *2018 26th European Signal Processing Conference (EUSIPCO)* (IEEE, Piscataway, 2018), pp. 1870–1874
36. D.E. Rumelhart, G.E. Hinton, R.J. Williams et al., Learning representations by back-propagating errors. Cognit. Model. **5**(3), 1 (1988)

Correction to: Data Analysis and Optimization for Engineering and Computing Problems

Pandian Vasant, Igor Litvinchev, Jose Antonio Marmolejo-Saucedo, Roman Rodriguez-Aguilar, and Felix Martinez-Rios

Correction to: P. Vasant et al. (eds.), *Data Analysis and Optimization for Engineering and Computing Problems*, EAI/Springer Innovations in Communication and Computing, https://doi.org/10.1007/978-3-030-48149-0

This book was inadvertently published without updating the following (or with the following error)

Affiliation of the Editor Roman Rodriguez-Aguila was erroneously published in the front matter, chapters 4, 5 and 14.

Affiliation to the chapter authors B. Retana-Blanco and E. Pedraza-Arroyo was erroneously published in chapter 4.

The above correction has now been updated.

The updated online versions of these chapters can be found at
https://doi.org/10.1007/978-3-030-48149-0_4
https://doi.org/10.1007/978-3-030-48149-0_5
https://doi.org/10.1007/978-3-030-48149-0_14
https://doi.org/10.1007/978-3-030-48149-0

P. Vasant et al. (eds.), *Data Analysis and Optimization for Engineering and Computing Problems*, EAI/Springer Innovations in Communication and Computing,
https://doi.org/10.1007/978-3-030-48149-0_21

Index

A
ABC method, 156–157, 162, 170
ADRMine, 220
Anesthesia, 27
ANFIS
  inventory policies, 169
  subsection sample, 160–161
  tool, 156
  using Matlab tool, 161
Asian energy ring, 181–182
  construction, 182
  North and Southeast Asia, 181
  northern ring, 181
Attributes
  and binary class, 208
  dependence range, 211
  minority class, 208
  PGMs, 208
  psychological, 218
  SMOTE-Cov, 208
Authentication
  AS, 145, 147
  and authorization, 144
  description, 145
  encryption and decryption of keys, 147
  Kerberos protocol, 144
  open network computing, 143
  private key, 147

B
Balancing conditions
  CLPB, 97
  computations, 101
  optimization problem, 87
  optimized packings, 95–106
  sparse clusters layout, 96
Basketball lineups, 62
Battery, 187
Bayesians networks, 157
Benders decomposition, 74, 75, 78

C
Cerebrospinal fluid (CSF)
  MRI image (*see* Magnetic resonance imaging (MRI))
  search and correction, 34
  segmentation algorithm, 36
  spinal cord and brain, 27
  volume, 28
Chang's algorithm, 112–115
Chronic diseases, 218–220, 222, 226
Cluster packing problem with balancing conditions (CLPB), 97
Computational results, 101–106
Covariance matrix
  best-known algorithms, 208
  datasets, 207
  experimental setup, 211–213
  IR, 207
  oversampling, 209–210
  PGMs, 208
  tools, 211–213

D
Decarburization
  China's energy plan, 178–180

P. Vasant et al. (eds.), *Data Analysis and Optimization for Engineering and Computing Problems*, EAI/Springer Innovations in Communication and Computing, https://doi.org/10.1007/978-3-030-48149-0

Decarburization (*cont.*)
GEIDCO roadmap
to 2035, 178
to 2050, 178
Decision making
ABC method, 156–157, 162
ANFIS, 160–161
Bayesians networks, 158
business efficiency, 127
EOQ, 157
fuzzy logic, 158–160
human reason, 136
industry 4.0, 126
inventory management, 155
organizations, 155
real-time data analysis, 75
safety stock, 157
SCM, 52
Didactic software
analysis and design, 112
design and layout, 118
distributed systems, 111
handling and manipulation, 117
IEEE 802.11 network protocols, 113
teaching tool prototype, 116
virtualized game, 112
WiFiSiM, 113
Diesel electrical generator (DGEN), 187
Digit pronunciation, 264
Distance optimization, 4, 43, 47, 85, 86, 96, 146
Distributed algorithms, 123
Distributed systems
election algorithms (*see* Teaching election algorithms)
Russia energy, 183
security risks, 143
teaching tools, 112
Drug opinions analysis
corpus processing, 221
learning process, 222
psycholinguistic feature extraction, 221–222, 225
Dynamic simulation, 52, 65

E
Economic order quantity (EOQ), 157
Eigenvalues, 197–204
Emergency medical services (EMS) organizations, 73–76
Emergency service, 76, 79
Energy ring
Asian, 181–182
European, 180
European energy ring, 180

F
Feasibility, 78, 125, 128, 175
Firefly algorithm
benchmark functions, 8
experimental design, 9–11
nature-inspired optimization, 3–4
parallel implementation, 5–7
rules, 4–5
First-grammar-then-voice (FGTV), 263, 264, 267, 268, 270
First-voice-then-grammar (FVTG), 263, 270–272
Fuzzy inference systems (FIS), 160, 161
Fuzzy logic
and experiments, 161
and FIS, 160
uncertainty, 157
vagueness, 158–160
variable processing
interface, 163
sets, 163
Takagi and Sugeno model, 166

G
Generalized singular decomposition (GVSD), 199–200
Global energy interconnection (GEI), 176–177
decarburization (*see* Decarburization)
GEIDCO, 190
shortage of energy, 190

H
Hadamard transform
example, 259–261
non-sinusoidal orthogonal, 256
sequence, 258–259
Hybrid model, 64
basketball lineups, 62
NBA team, 61
optimization, 62–63
performance lineups, 65–69
players' performance, 62
stochastic frontiers, 63–64
technical efficiency, 63–64

I

Image registration, 29, 34
Imbalanced datasets, 207, 211
Imbalance ratio (IR), 207, 209
Industry 4.0 technologies, 125, 126, 130
  automation, 136
  big data, 137
  collaborative robotics, 137
  load sensing, 137
  M2M, 137
  predictive maintenance, 137
  simulation, 136
Information theory, 244
Intralogistics, 134–135
  Colombian companies, 126
  concepts, 127
  cyberphysical systems, 127
  industry 4.0, 126
  industry and technologies 4.0, 136
  methodology
    choice of categories, 132–133
    data compilation, 129–131
    material evaluation, 133
    sampling planning, 131–132
  problems and challenges, 126
  simulation model, 128
  storage loading, 135
Intralogistics 4.0, 127, 134, 138
Inventory control
  ANFIS, 169
  output defuzzification, 168–169
  rules, 167

K

Kerberos
  authentication protocol, 143
  Diffie-Hellman-DSA, 144
  proposed key distribution, 149
  protocol, 144
  servers, 145
Key distribution center (KDC), 144, 145, 245
Key distribution prototype
  architecture, 145–148
  evaluation and analysis, 152–153
  implementation
    AES, 149
    MySQL, 151, 152
    RSA, 150–151
    servers, 152
  Kerberos authentication, 143
  related work, 144–145
  TICs, 143

L

Layout problem
  mathematical model, 86–89
  problem formulation, 86–89
  science and technology, 85
  shock wave processes, 86
  solution strategy, 89–91
Linear optimization, 197, 198
Location
  ambulance bases, 82
  analysis
    facility, 55
    historical data, 54
    population index, 54–55
    products, 56
  bilevel model, 74
  case study, 79–80
  citizen assistance, 81
  emergency medical systems, 75
  EMS, 73
  notation, 76–78
  road traffic injuries, 74
  solution method, 78
  traffic accidents, 74
Logistics operators, 125–129, 132, 133, 138, 139

M

Magnetic resonance imaging (MRI)
  connected component, 30
  CSF, 28
  evaluation of results, 36
  materials and computational methodology
    algorithm, 32–33
    computational complexity, 34
    pattern recognition, 30
    proposed algorithm, 36
    spine, 31
    transverse view, 32
    type T1 and T2, 30
  metrics, 35–36
  planes/types of, 28
  radiofrequency, 28
  registration of images, 29
  segmentations, 34–35
  types of images, 29, 30
Mamdani model, 164–166, 168
Mathematical model, 47, 62, 98–99
  equilibrium packing clusters, 106
  problem formulation, 86–89
  TM, 232
Metaheuristics, 13, 14, 18, 92

Mexican automotive industry
assignment dealers automotive company, 54, 59
data analysis
additional information, 46–47
assignment of cases, 44, 46
percentage of allocation, sales, 44, 45
sales forecast, 46–49
database headers, 42
dealers automotive company, 54, 59
information, 42
literature review, 42–44
mathematical model, 47
percentage of allocation per vehicle, 50–51, 56–59
software used, 52–54
Mixed PCA application
Eigenvalues, 200, 202–204
ENIGH 2016, 200, 201
rotation, 204, 205
Multiple correspondence analysis (MCA), 196, 199, 200, 203, 205
Multithreading, 13
Multivariate analysis, 195

N
Nature-inspired algorithms, 3–4, 13, 14
NBA team, 61, 65, 70
Network coding
butterfly topology, 243, 244
connectivity schemes, 244
evaluation, 246–252
implementation, 246–252
information transmission schemes, 243
methodology, 246
pollution attack, 244
related work, 245–246
source nodes, 243
Network optimization, 47, 244
Neural networks
ANN, 161, 219
and learning algorithm, 264, 267–268
pattern recognition, 266
recursive, 264
short-time Fourier transform, 272
Nonlinear programing, 92, 96, 98

O
Oversampling
AUC, data balancing algorithms, 212
covariance matrix, 209
CovI and CovO variants, 212
SMOTE-Cov, 210

P
Packing
computational results, 101–106
mathematical modeling, 98–99
NP-hard, 95
objects and containers, 95
problem statement, 96–98
solution algorithm, 99–101
Parallel algorithm
firefly, 6–7
multicore processors, 5–6
restarting techniques, 7
Parallel implementation
firefly algorithm, 5–7
genetic restart to the best, 20, 21
restart to the best, 18–20
Pattern recognition, 30, 264, 266
Pollution attacks
communication scheme, 246
data coding, 245
inefficient communication, 244
node N1 and R1, 249
node Z, 251
Power plants, 176, 181, 183–184, 190
Prehospital services, 73, 79
Principal component analysis (PCA)
exploratory data analysis, 195–196
generalized singular value decomposition, 199–200
mixed application (*see* Mixed PCA application)
standard, 196–198
Pseudo-random generators, 257, 262
Psycholinguistic features, 218, 220–222, 225, 226

R
Rainfall optimization algorithm (RFO)
benchmark functions, 15
description, 16
experimental results, 21–22
explosion process, 17
merit-order list, 16
multivariable functions, 14
nature-inspired optimization, 13
parallel implementation, 18–21
steepest slopes, 14, 16
Reduction technique, 233, 242

Region growing, 33
Renewable energy sources (RES)
  Asian energy ring, 181–182
  European energy ring, 180
  and GEI, 176–177
  Gobitec and Asia network projects, 182
  high-efficiency characteristics, 175
  human development, 175
  in Myanmar, 184–189
  optimization model, 184, 185
  power
    plants, 183–184
    system with installations, 184, 185
  security, 175
Rivest, Shamir, Adleman (RSA), 144–147, 150–152

S
Safety stock, 157, 170
Security
  encryption/decryption, 150
  exchange keys, 144
  implementation, 144
  network coding schemes (*see* Network coding)
  and privacy issues, 217
  RES, 182
  risks, 143
  threat detection, 207
  TICs, 143
Segmentation, 28, 29, 34–36
Sentiment analysis method
  drug opinions analysis, 220–222
  evaluation
    corpus, 224
    results, 224–225
    subjects, 222–224
  on-going and fast-growing subfield, 218
  related work, 219–220
  security and privacy issues, 217
Sentiment polarity classification, 219
Simulation, 52, 113, 122, 123, 126, 128, 131, 136–137
SMOTE-Cov
  AUC, 212
  balance datasets, 210
  covariance matrix, 210
  experimental study, 208
Solar energy, 178, 187, 190
Solution algorithm, 99–101
Speech processing
  audio signals, 265–266
  contribution, 264
  FGTV, 268–270
  FVTG, 270–272
  learning algorithm, 267–268
  motivation, 263
  neural network, 267–268
  NLP, 264
  organization, 264–265
  preprocessing, 266–267
  problem statement, 263
  state of the art, 263–264
Sports analytics, 62
Steganalysis
  proofs, 237–240
  research, 229
  and steganography techniques, 233–234
  and TM, 233–234
  undecidability, 231, 234–237
  universal tool, 234–237
Steganography
  data compression, 236
  definition, 230–231
  TM, 232
  undecidability, 232
Stochastic frontiers, 63–64, 67, 68, 70
Stochastic programming, 75, 78
Storage loading, 135–136
  categories, 128
  and intralogistics (*see* Intralogistics)
  organizational efficiency, 139
  storage tool, 128
  and unloading, 126
Stream ciphers
  auto-correlations, 255
  cryptanalysis, 255
  example, 259–261
  sequence, 258–259
Supply chain
  dealers automotive company, 52, 53
  distribution centers, 43
  inventory classification, 156
  logistics, 131
  requirements, 128
  SCM, 52
Supply chain management (SCM), 43, 52, 155
Swarm intelligence, 4

T
Takagi and Sugeno model, 167
Teaching election algorithms
  Chang's algorithm, 113–114
  digital revolution, 111
  evaluation, 120–123
  implementation, 118–120

Teaching election algorithms (*cont.*)
interface prototypes, 116–117
object repository design, 117–118
related work, 112–113
requirements analysis, 114–115
tool structure, 112, 115
Technical efficiency
Boston lineups, 69
indicator, 68, 70
model, 63–64
net rating, 70
performance, 67
Technologies 4.0, 125, 126, 130
automation, 137
big data, 137
collaborative robotics, 137
load sensing, 137
M2M, 137
predictive maintenance, 137
simulation, 136–137
Test dataset, 219
Turing machine (TM)
cover image, 236
definition, 232
diagonalization technique, 239
hidden information, 238
steganalysis, 236
steganographic algorithm, 233–234

U
Undecidability
proofs, 237–240
steganalysis techniques, 231
steganography, 233–234
TM, 233–234
types of problems, 232
universal steganalysis tool, 234–237
Unified energy system (UES) of Russia, 184, 185
Unified power system, 183–184

V
Variable processing
defuzzification, 164
fuzzy system
interface, 163
logic, 162–163
sets, 163
Mamdani model, 164–166
model Mamdani *vs.* Takagi-Sugeno (TS), 167
Takagi and Sugeno model, 166–167
Vulnerability, 207, 245

W
Wind energy, 178, 184, 186, 187

The manufacturer's authorised representative in the EU is Springer Nature Customer Service Centre GmbH, Europaplatz 3, 69115 Heidelberg, Germany. If you have any concerns regarding our products, please contact ProductSafety@springernature.com

Printed and bound by CPI Group (UK) Ltd, Croydon, CR0 4YY
22/09/2026
02221060-0010